McGraw-Hill
My Math

This is your very own math book! You can write in it, draw, circle, and color as you explore the exciting world of math.

Let's get started. Grab a crayon and draw a picture that shows what math means to you.

Have fun!

This is your space to draw.

McGraw Hill **Education**

Bothell, WA • Chicago, IL • Columbus, OH • New York, NY

connectED.mcgraw-hill.com

Education

STEM McGraw-Hill is committed to providing
instructional materials in Science, Technology, Engineering,
and Mathematics (STEM) that give all students a solid
foundation, one that prepares them for college and careers
in the 21st century.

Send all inquiries to:
McGraw-Hill Education
STEM Learning Solutions Center
8787 Orion Place
Columbus, OH 43240

ISBN: 978-0-02-116067-9 (Volume 2)
MHID: 0-02-116067-8

Printed in the United States of America.

20 LWI 19 18 17

Our mission is to provide educational resources that enable
students to become the problem solvers of the 21st century
and inspire them to explore careers within Science, Technology,
Engineering, and Mathematics (STEM) related fields.

Meet The Artists!

Marya Barnaba
Lauren Schonowoski

Shape Friends=Math Friends Math and Art work together like good friends. Art makes learning math fun. Patterns, shapes and colors and adding a little imagination become Shape Friends. *Volume 1*

Levin Neighbors

Carpet Store Math Math is cool! My Daddy uses a lot of math at the carpet store! *Volume 2*

Other Finalists

Eric Fernandez
Primary Color Addition 3

Alyssa Sullivan
Dress Up With Numbers

Chloe Uemura
Butterfly Fairy

Natalie Rush
Sculpted Patterns

Trinity Williams
Math is Fun by Trinity

Xia Peterson
Get Mathy

*Lisa Hart's Class**
Our Math in Our Hands

*Sharon E. Davison's Class**
Math is Lots of Things to Us

Samantha Carrington
Lily

Find out more about the winners and other finalists at www.MHEonline.com.

We wish to congratulate all of the entries in the 2011 *McGraw-Hill My Math* "What Math Means To Me" cover art contest. With over 2,400 entries and more than 20,000 community votes cast, the names mentioned above represent the two winners and nine finalists for this grade.

* Please visit mhmymath.com for a complete list of students who contributed to this artwork.

GO digital

it's all at
connectED.mcgraw-hill.com

Go to the Student Center for your eBook, Resources, Homework, and Messages.

Write your Username [_____]   Password [_____]

Get your resources online to help you in class and at home.

Vocab

Find activities for building vocabulary.

Watch

Watch animations of key concepts.

Tools

Explore concepts with virtual manipulatives.

Check

Self-assess your progress.

eHelp

Get targeted homework help.

Games

Reinforce with games and apps.

Tutor

See a teacher illustrate examples and problems.

GO mobile

Scan this QR code with your smart phone* or visit mheonline.com/stem_apps.

*May require quick response code reader app.

Available on the **App Store**

Contents in Brief
Organized by Domain

Common Core State Standards

Standards for Mathematical PRACTICE → Woven Throughout

Chapter

Numbers 0 to 5

Getting Started

Quack!
Quack!

Lessons and Homework

Wrap Up

connectED.mcgraw-hill.com

Chapter 2 Numbers to 10

ESSENTIAL QUESTION
What do numbers tell me?

Getting Started

Lessons and Homework

Wrap Up

connectED.mcgraw-hill.com

We like healthful food!

Chapter 3 Numbers Beyond 10

ESSENTIAL QUESTION
How can I show numbers beyond 10?

Getting Started

Blast off!

Lessons and Homework

Wrap Up

connectED.mcgraw-hill.com

Chapter 4 Compose and Decompose Numbers to 10

ESSENTIAL QUESTION
How can we show a
number in other ways?

Getting Started

Lessons and Homework

Wrap Up

connectED.mcgraw-hill.com

Chapter 5 Addition

ESSENTIAL QUESTION
How can I use objects to add?

Getting Started

Lessons and Homework

Wrap Up

Party time!

connectED.mcgraw-hill.com

Chapter 6 Subtraction

Fitness fun!

connectED.mcgraw-hill.com

Chapter 7 Compose and Decompose Numbers 11 to 19

ESSENTIAL QUESTION
How do we show numbers 11 to 19 in another way?

Getting Started

Lessons and Homework

Wrap Up

connectED.mcgraw-hill.com

Let it snow!

Chapter

8 Measurement

Getting Started

Lessons and Homework

Fun in the sun!

Wrap Up

connectED.mcgraw-hill.com

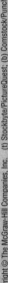

Chapter 9 Classify Objects

ESSENTIAL QUESTION
How do I sort objects?

Getting Started

Lessons and Homework

Wrap Up

connectED.mcgraw-hill.com

Bright idea!

WE RECYCLE

Chapter 10 Position

ESSENTIAL QUESTION
How do I identify positions?

Getting Started

Lessons and Homework

Wrap Up

Animals in action!

connectED.mcgraw-hill.com

Chapter 11 Two-Dimensional Shapes

Getting Started

Lessons and Homework

Wrap Up

connectED.mcgraw-hill.com

Let's learn shapes!

Chapter 12 Three-Dimensional Shapes

Getting Started

Lessons and Homework

Wrap Up

connectED.mcgraw-hill.com

Shapes are fun!

Chapter

8 Measurement

ESSENTIAL QUESTION

How do I describe and compare objects by length, height, and weight?

Let's Go Camping!

Watch a video!

Watch ▶

My Common Core State Standards

Measurement and Data

K.MD.1 Describe measurable attributes of objects, such as length or weight. Describe several measurable attributes of a single object.

K.MD.2 Directly compare two objects with a measurable attribute in common, to see which object has "more of"/"less of" the attribute, and describe the difference.

Standards for Mathematical PRACTICE ⬇

1. Make sense of problems and persevere in solving them.
2. Reason abstractly and quantitatively.
3. Construct viable arguments and critique the reasoning of others.
4. Model with mathematics.
5. Use appropriate tools strategically.
6. Attend to precision.
7. Look for and make use of structure.
8. Look for and express regularity in repeated reasoning.

= focused on in this chapter

Name

 Check ← Go online to take the Readiness Quiz

1

2

3

4

 Directions: 1–2. Circle the object that is different. **3.** Circle the object that is longer. **4.** Circle the animal that you could hold in your hand.

My Math Words

Review Vocabulary

bigger   smaller

Directions: Circle the bigger duck. Draw an X on the smaller ducks. Trace each word.
Draw a bigger frog next to the frog.

My Vocabulary Cards

Vocab abc

capacity

heavier

↑ heavier

height

holds less

↑ holds less

holds more

↑ holds more

length

Teacher Directions:
Ideas for Use

- Have students choose two objects in the classroom. Have students choose a vocabulary card and use the vocabulary word to compare the two objects.

- Have students name the letters in each word.

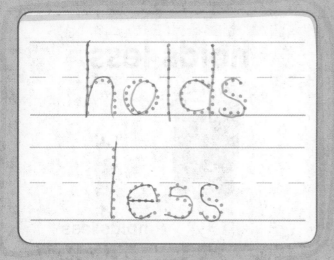

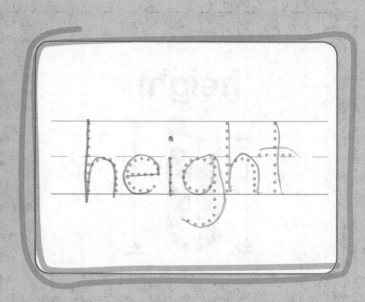

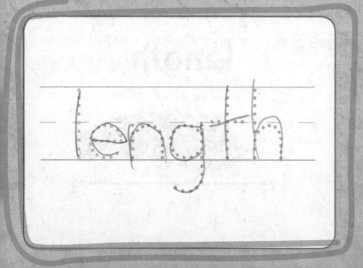

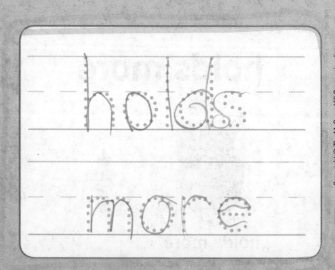

lighter

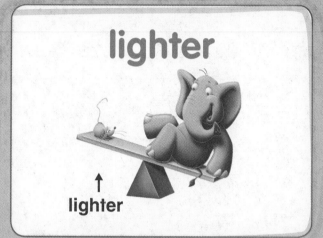

lighter

longer

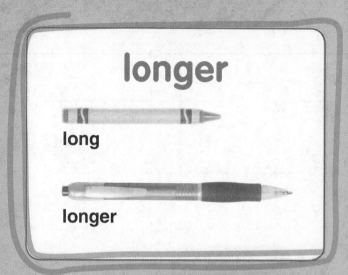

long

longer

shorter

short   shorter

taller

tall   taller

weight

Teacher Directions:
More Ideas for Use

• Tell students to create riddles for each word. Ask them to work with a friend to guess the word for each riddle.

• Help students use the blank card to write a word from a previous chapter that they would like to review.

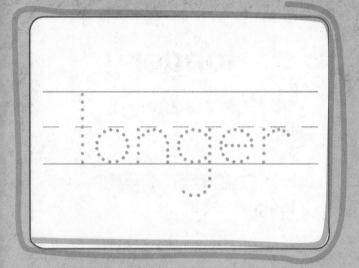

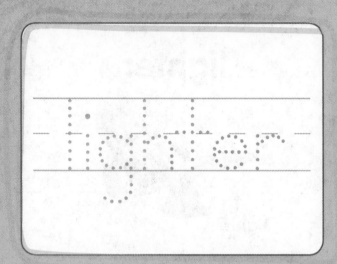

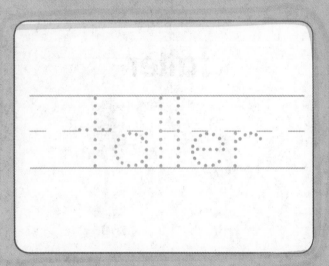

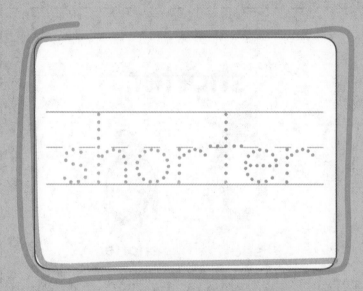

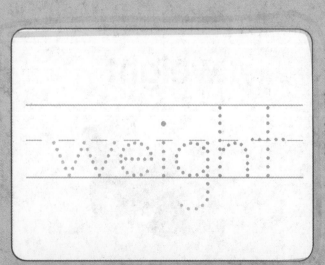

longer

shorter

shorter

taller

heavier

lighter

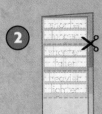
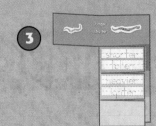

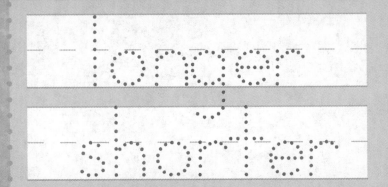

longer

shorter

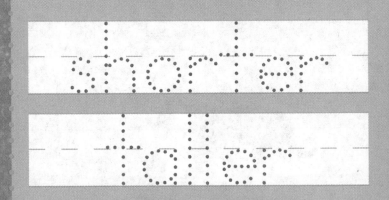

shorter

taller

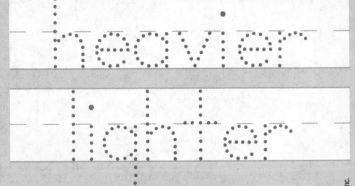

heavier

lighter

Measurement and Data
K.MD.1, K.MD.2

CCSS

Compare Length

Lesson 1

ESSENTIAL QUESTION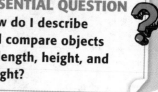
How do I describe and compare objects by length, height, and weight?

Explore and Explain Tools Watch

Go fish!

 Teacher Directions: Use to make a train shorter than the fish. Draw the train above the fish. Use cubes to make a train longer than the fish. Draw the train below the fish.

Online Content at ⟡ **connectED.mcgraw-hill.com**

See and Show

1 length   longer   shorter

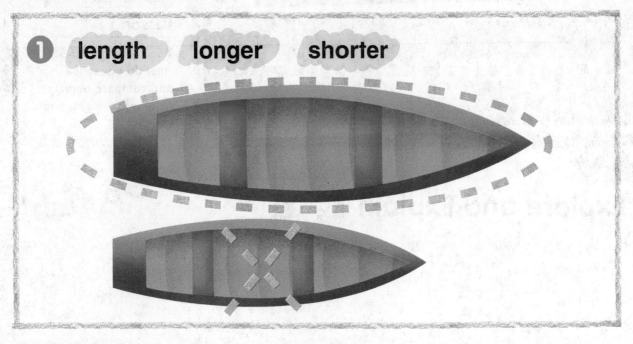

2

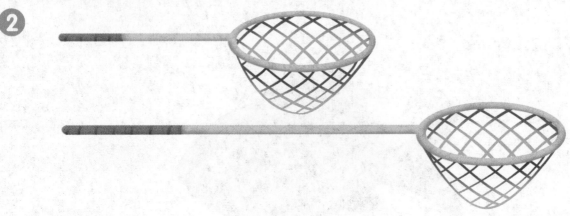

3

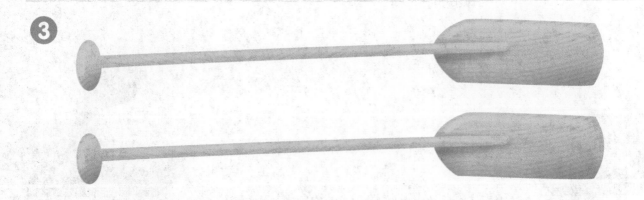

Directions: 1. Compare the objects. Trace the X on the object that is shorter. Trace the circle around the object that is longer. **2–3.** Compare the objects. Draw an X on the object that is shorter. Draw a circle around the object that is longer. If the objects are the same length, underline them.

Name

....

On My Own

4

5

6

7

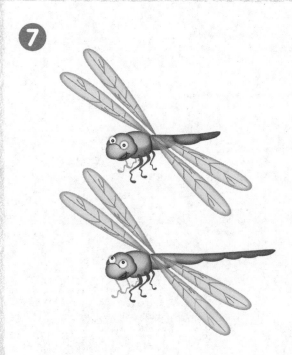

 Directions: 4–7. Compare the objects. Draw an X on the object that is shorter. Draw a circle around the object that is longer. If the objects are the same length, underline them.

Online Content at ⬏ **connectED.mcgraw-hill.com** Chapter 8 • Lesson 1   491

Copyright © The McGraw-Hill Companies, Inc. Photodisc/Getty Images

8

Directions: 8. Use connecting cubes to make a train the same length as the snake. Use cubes to make a train shorter and a train longer than the snake. Draw a shorter snake above the snake and a longer snake below the snake.

Name

My Homework

Lesson 1

Compare Length

Homework Helper

eHelp

Need help? connectED.mcgraw-hill.com

1

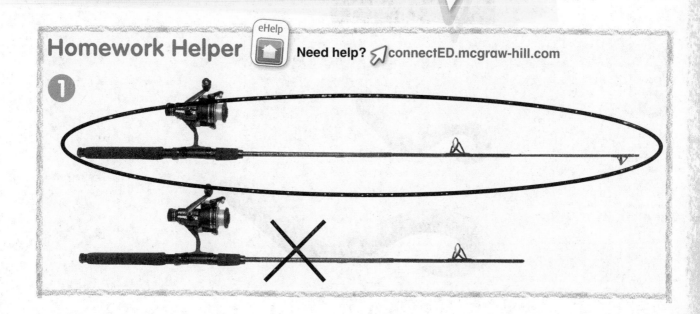

2

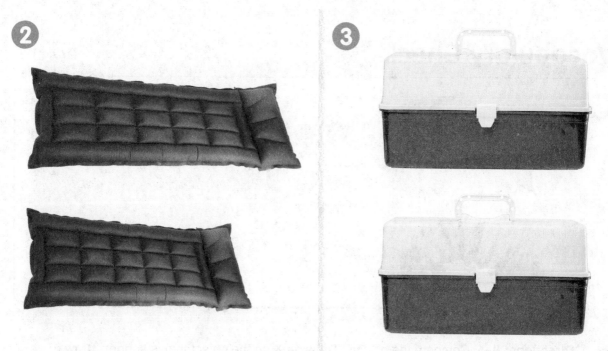

3

Directions: 1–3. Compare the objects. Draw an X on the object that is shorter. Draw a circle around the object that is longer. If the objects are the same length, underline them.

4

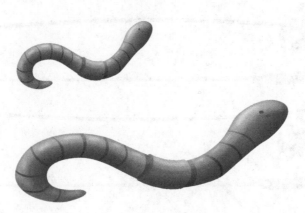

5

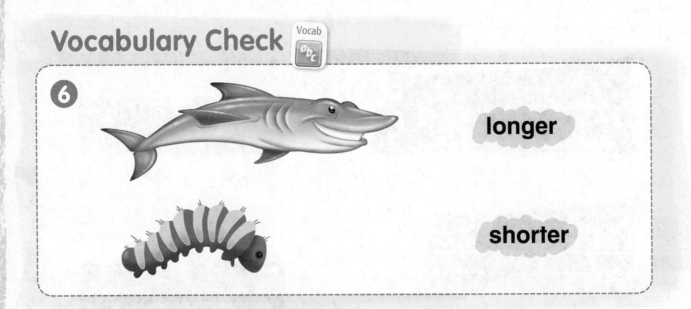

Vocabulary Check

6

longer

shorter

Directions: 4–5. Compare the objects. Draw an X on the object that is shorter. Draw a circle around the object that is longer. **6.** Draw a line from the word *longer* to the longer animal. Draw a line from the word *shorter* to the shorter animal.

Math at Home Place a spoon and pencil on the table. Have your child tell which is shorter, which is longer, or if the objects are the same length.

Name _____

Compare Height

Lesson 2

ESSENTIAL QUESTION
How do I describe and compare objects by length, height, and weight?

Explore and Explain

Tools Watch

To the moon!

 Teacher Directions: Use ▬ to show a rocket that is taller than the rocket on the page. Trace the connecting cubes.

See and Show

1 height

shorter

taller

2

3

Directions: 1. Compare the objects. Trace the X on the object that is shorter. Trace the circle around the object that is taller. **2–3.** Compare the objects. Draw an X on the object that is shorter. Draw a circle around the object that is taller. If the objects are the same height, underline them.

Name

On My Own

4

5

6

7

 Directions: 4–7. Compare the objects. Draw an X on the object that is shorter. Draw a circle around the object that is taller. If the objects are the same height, underline them.

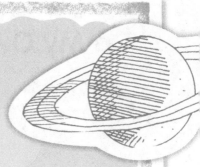

8

Directions: 8. Draw a tall object. Draw a short object beside it. Have a classmate tell which object is short and which object is tall.

Name

My Homework

Lesson 2

Compare Height

Homework Helper

eHelp

Need help? connectED.mcgraw-hill.com

1

2

3

Directions: 1–3. Compare the objects. Draw an X on the object that is shorter. Draw a circle around the object that is taller. If the objects are the same height, underline them.

④

⑤

Vocabulary Check

⑥

taller　　　**shorter**

Directions: 4–5. Compare the objects. Draw an X on the object that is shorter. Draw a circle around the object that is taller. **6.** Draw a line from the word *taller* to the taller animal. Draw a line from the word *shorter* to the shorter animal.

Math at Home Gather a pencil and a crayon. Stand them side by side. Ask your child to tell which is shorter, which is taller, or if they are the same height. Compare other objects.

Name ....

Problem Solving
STRATEGY: Guess, Check, and Revise

Lesson 3

ESSENTIAL QUESTION
How do I describe and compare objects by length, height, and weight?

How long?

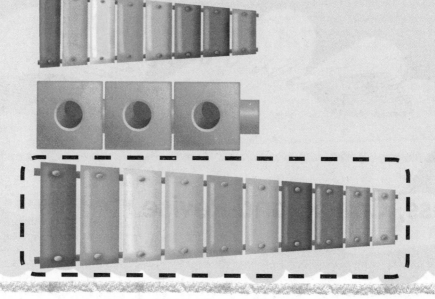

Guess, Check, and Revise

guess   check

Teacher Directions: Compare the objects. Trace the circle around the object that is longer. Then guess how many cubes long the longer object is. Write the number. Use cubes to check. Is your answer close? Trace the cubes. Trace the number.

How long?

Guess, Check, and Revise

_____ _____

- - - - - - - - - - - -

_____ _____

guess   check

Directions: Compare the objects. Circle the object that is longer. Then guess how many cubes long the longer object is. Write the number. Use cubes to check. Is your answer close? Trace the cubes. Write the number.

Name _____

How long?

Guess, Check, and Revise

_____ _____

_____ _____

_____ _____

guess   check

 Directions: Compare the objects. Circle the object that is longer. Then guess how many cubes long the longer object is. Write the number. Use cubes to check. Is your answer close? Trace the cubes. Write the number.

How long?

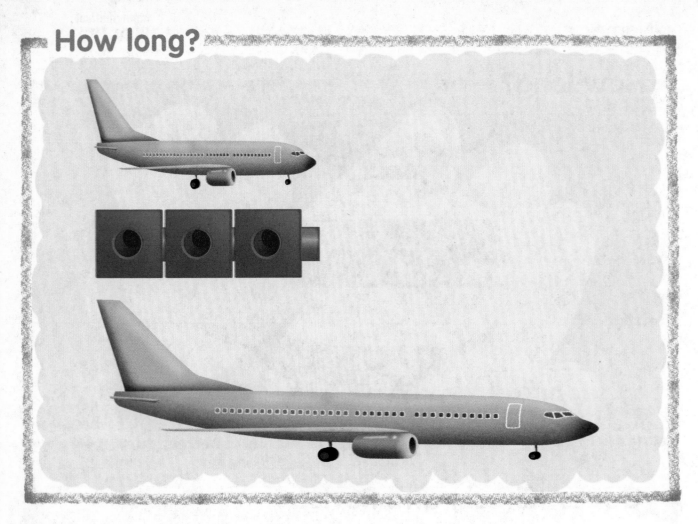

Guess, Check, and Revise

_____ _____

_____ _____

_____ _____

guess   check

Directions: Compare the objects. Circle the object that is longer. Then guess how many cubes long the longer object is. Write the number. Use cubes to check. Is your answer close? Trace the cubes. Write the number.

Measurement and Data
K.MD.1, K.MD.2

CCSS

My Homework

Lesson 3

Problem Solving: Guess, Check, and Revise

How long?

ADMISSION TICKET

Admit One

ADMISSION TICKET

Admit One

Guess, Check, and Revise

guess   check

Directions: Compare the objects. Circle the object that is longer. Then guess how many pennies long the longer object is. Write the number. Use pennies to check. Is your answer close? Trace the pennies. Trace the number.

How long?

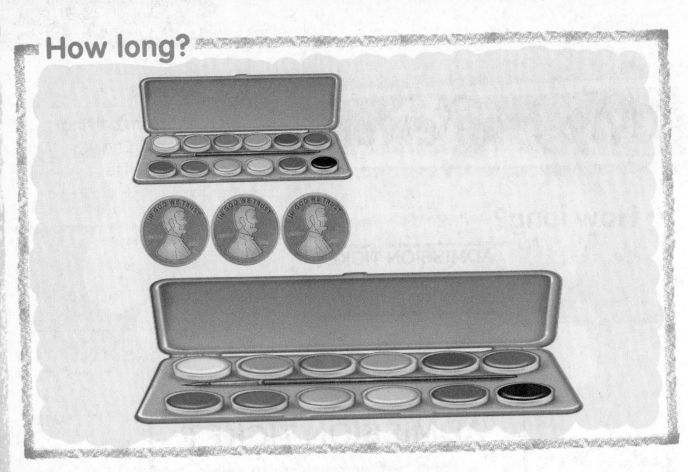

Guess, Check, and Revise

guess   check

Directions: Compare the objects. Circle the object that is longer. Then guess how many pennies long the longer object is. Write the number. Use pennies to check. Is your answer close? Trace the pennies. Write the number.

Math at Home Take advantage of problem-solving opportunities during daily routines such as riding in the car, bedtime, doing laundry, putting away groceries, and so on.

Name

Vocabulary Check

1 **longer**

shorter

2

shorter   **taller**

Concept Check

3

 Directions: 1. Look at the pencil. Draw a pencil that is shorter. **2.** Look at the cup. Draw a cup that is taller. **3.** Compare the objects. Draw an X on the object that is shorter. Draw a circle around the object that is taller.

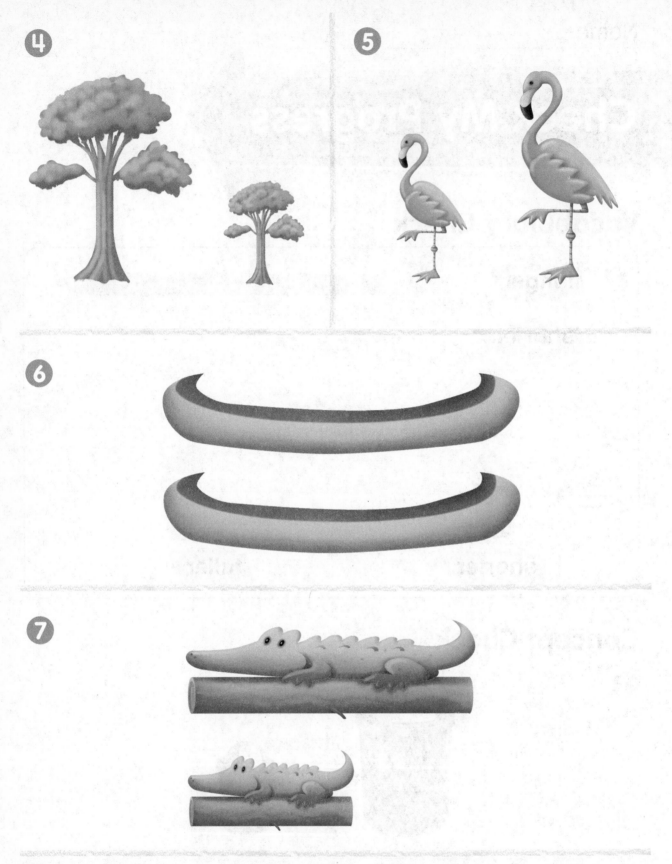

Directions: 4–5. Compare the objects. Draw an X on the object that is shorter. Draw a circle around the object that is taller. **6–7.** Compare the objects. Draw an X on the object that is shorter. Draw a circle around the object that is longer. If the objects are the same length, underline them.

Name

Compare Weight

Lesson 4

ESSENTIAL QUESTION
How do I describe and compare objects by length, height, and weight?

Hi!

Explore and Explain

 Teacher Directions: Find two different objects in the classroom. Tell which object is heavy and which is light. Draw a picture of each object on the teeter-totter to show which object is heavy and which is light.

See and Show

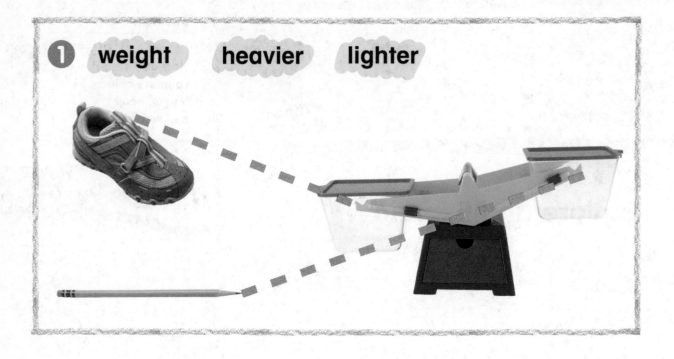

1 weight   heavier   lighter

2

3

Directions: 1. Compare the objects. Trace the line from each object to the place on the balance scale that shows its weight. **2–3.** Compare the objects. Circle the heavier object. Draw an X on the lighter object. If the objects weigh the same, underline them.

Name

On My Own

This ball is light!

4

5

6

Directions: 4. Compare the objects. Draw a line from each object to the place on the balance scale that shows its weight. **5–6.** Compare the objects. Circle the heavier object. Draw an X on the lighter object. If the objects weigh the same, underline them.

Real World **Problem Solving**

7

 Directions: 7. Draw an X on items that are too heavy to lift. Circle the items that are light enough to carry.

Name _____

Measurement and Data
K.MD.1, K.MD.2

CCSS

My Homework

Lesson 4

Compare Weight

Homework Helper

Need help? connectED.mcgraw-hill.com

1

2

3

 Directions: 1–3. Compare the objects. Circle the heavier object. Draw an X on the lighter object. If the objects weigh the same, underline them.

4

5

Vocabulary Check

6   lighter   heavier

Directions: 4–5. Compare the objects. Circle the heavier object. Draw an X on the lighter object. If the objects weigh the same, underline them. **6.** Draw a line from the word *lighter* to the lighter object. Draw a line from the word *heavier* to the heavier object.

Math at Home   Use a canned good and an empty cup. Place one item in each of your child's hands. Ask your child which hand holds the heavier item and which hand holds the lighter item.

Name ....

Describe Length, Height, and Weight

Explore and Explain

Lesson 5

ESSENTIAL QUESTION
How do I describe and compare objects by length, height, and weight?

Which one is longer?

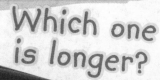

Which one is heavier?

Teacher Directions: Choose two real-world objects. Draw them. Describe the length of each object. Circle the object that is longer. Describe the weight of each object. Draw an X on the object that weighs more.

See and Show

Directions: 1–2. Look at the objects. Describe the length of each object. Circle the object that is longer. Describe the weight of each object. Draw an X on the object that weighs more. **3–4.** Look at the objects. Describe the height of each object. Circle the object that is taller. Describe the length of each object. Draw an X on the object that is longer.

Name

On My Own

This ball is heavy!

 5

 Glue Stick

 6

 7

8

 Directions: 5–8. Look at the objects. Describe the height of each object. Circle the object that is taller. Describe the weight of each object. Draw an X on the object that weighs more.

Problem Solving

 9

 10

Directions: 9. Look at the box of bandages. Describe the objects in the row. Circle the object that is longer and weighs more than the box of bandages.   **10.** Look at the bowl. Describe the objects in the row. Circle the object that is taller and weighs more than the bowl.

Name

My Homework

Lesson 5

Describe, Length, Height, and Weight

Homework Helper

eHelp

Need help? connectED.mcgraw-hill.com

1

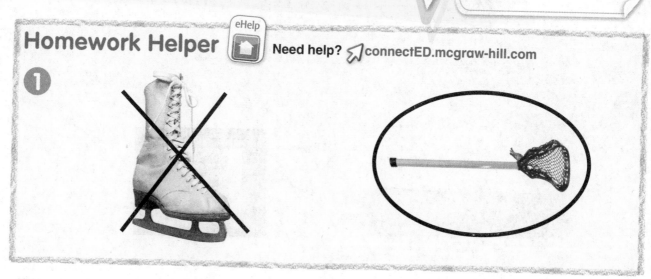

2

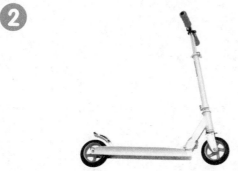

3

Directions: 1. Look at the objects. Describe the length of each object. Circle the object that is longer. Describe the weight of each object. Draw an X on the object that weighs more. **2–3.** Look at the objects. Describe the height of each object. Circle the object that is taller. Describe the length of each object. Draw an X on the object that is longer.

❹

❺

❻

Directions: 4. Look at the objects. Describe the height of each object. Circle the object that is taller. Describe the weight of each object. Draw an X on the object that weighs more. **5–6.** Look at the objects. Describe the length of each object. Circle the object that is longer. Describe the weight of each object. Draw an X on the object that weighs more.

Math at Home Gather items in your house. Ask your child to tell which is taller. Ask your child to tell which is longer. Ask your child to tell which item weighs more.

Name ....

Measurement and Data
K.MD.1, K.MD.2

CCSS

Compare Capacity

Lesson 6

ESSENTIAL QUESTION
How do I describe and compare objects by length, height, and weight?

Explore and Explain

Tools   Watch

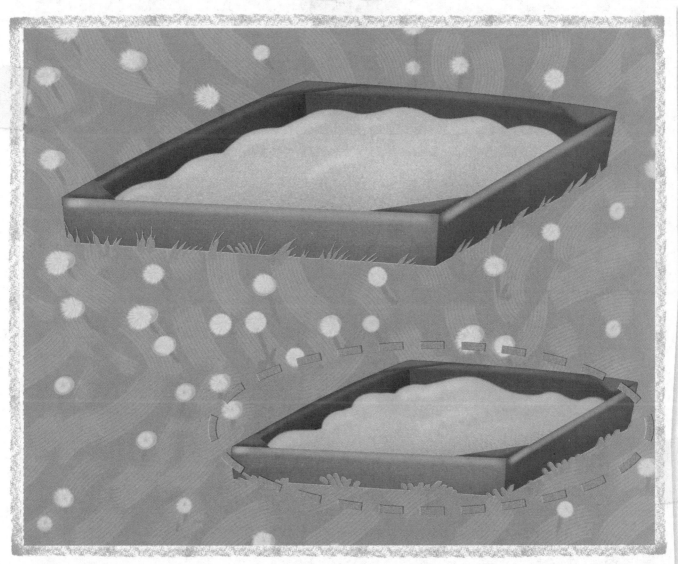

 Teacher Directions: Use ▮ to fill in both sand boxes on the page. Tell which sand box holds more. Circle the sand box that holds less.

See and Show

1

capacity

holds more

holds less

2

3

Directions: 1. Compare the containers. Trace the circle around the object that holds more. Trace the X on the object that holds less. **2–3.** Compare the containers. Circle the object that holds more. Draw an X on the object that holds less. If the objects hold the same, underline them.

Name _____

On My Own

 4

 5

6

 Directions: 4–6. Compare the containers. Circle the object that holds more. Draw an X on the object that holds less. If the objects hold the same, underline them.

Problem Solving

Which one holds more?

7

Directions: 7. Look at the object in the box. Draw an X on items that hold more than the object. Circle the items that hold less than the object.

Name

My Homework

Lesson 6
Compare Capacity

Homework Helper  Need help? connectED.mcgraw-hill.com

1

2

3

 Directions: 1–3. Compare the containers. Circle the object that holds more. Draw an X on the object that holds less. If the objects hold the same, underline them.

Chapter 8 • Lesson 6   525

4

5

Vocabulary Check

6

holds more　　　**holds less**

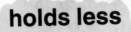

 Directions: 4–5. Compare the containers. Circle the object that holds more. Draw an X on the object that holds less. If the objects hold the same, underline them. **6.** Draw a line from the object that holds more to the words *holds more*. Draw a line from the object that holds less to the words *holds less*.

Math at Home Use a sauce pan and empty cup. Ask your child which object holds more and which object holds less.

Name

Vocabulary Check

taller

shorter

holds less

holds more

heavier

lighter

 Directions: 1. Circle the tent that is taller. Draw an X on the tent that is shorter.
2. Circle the cooler that holds more. Draw an X on the cooler that holds less.
3. Circle the bag that is heavier. Draw an X on the bag that is lighter.

Concept Check

1

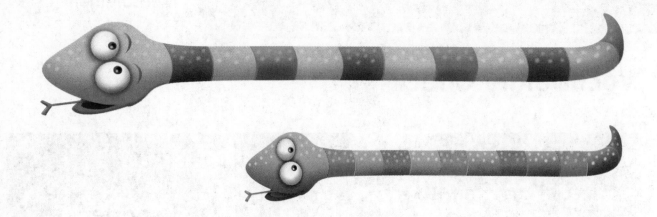

2

3

Directions: 1. Compare the objects. Draw an X on the object that is shorter. Circle the object that is longer. **2.** Draw an X on the object that is heavier. Circle the object that is lighter. **3.** Draw an X on the object that holds more. Circle the object that holds less.

Problem Solving

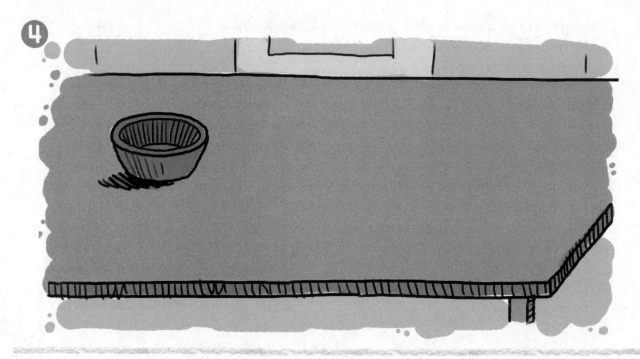

 Directions: 4. Look at the bowl. Draw a bowl that holds more next to it.
5. Look at the tent. Draw a shorter tent next to it.

Reflect

Chapter 8

ESSENTIAL QUESTION
How do I describe and compare objects by length, height, and weight?

1

2

 Directions: 1. Draw one long object. Draw a shorter object below it. **2.** Draw a tall object. Draw a shorter object beside it. Have a classmate tell which object is short and which object is tall.

Chapter

 Classify Objects

Earth Needs Our Help!

Watch

Watch a video!

My Common Core State Standards

Measurement and Data

K.MD.3 Classify objects into given categories; count the numbers of objects in each category and sort the categories by count.

Standards for

Mathematical PRACTICE

1. Make sense of problems and persevere in solving them.
2. Reason abstractly and quantitatively.
3. Construct viable arguments and critique the reasoning of others.
4. Model with mathematics.
5. Use appropriate tools strategically.
6. Attend to precision.
7. Look for and make use of structure.
8. Look for and express regularity in repeated reasoning.

= focused on in this chapter

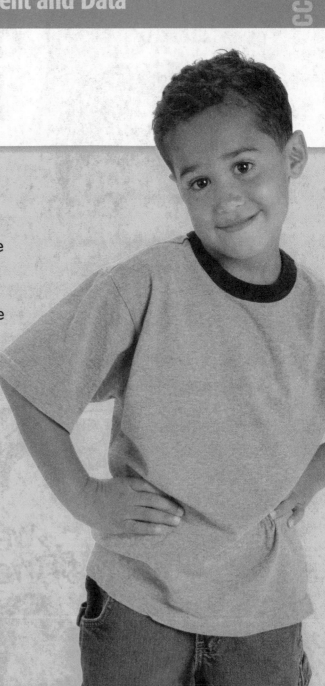

Name

Am I Ready?

← Go online to take the Readiness Quiz

1

2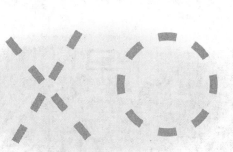

3

4

 Directions: 1. Trace the dashed marks. **2.** Circle the car. Draw an X on the tree.
3. Color the truck red. Color the ball blue. Color the bird yellow.
Color the flower purple. **4.** Circle the object that is small.

My Math Words

Review Vocabulary

Directions: Trace each word. Draw lines to match each word with the recycling bins that show that size.

alike

different

shape

size

sort

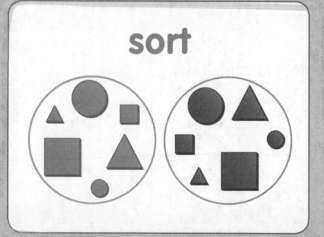

Teacher Directions:
Ideas for Use

- Have students choose two words. Have them tell if any of the letters in the words are alike. Have them point to letters that are different.

- Have students name the letters in each word.
- Have students use the blank card to write their own vocabulary word.

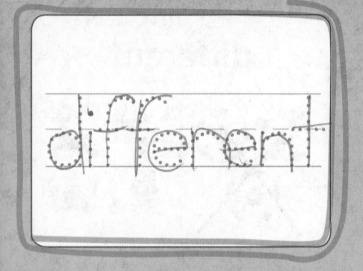

different

alike

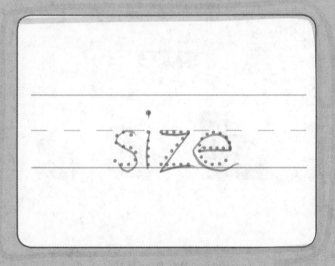

size

shape

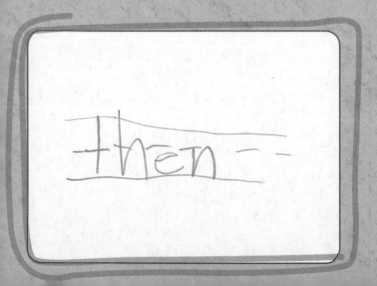

then

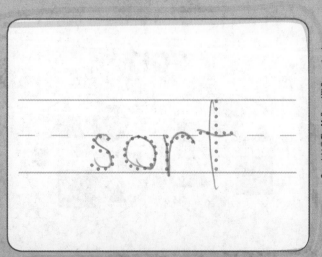

sort

Name ....

Alike and Different

Lesson 1

ESSENTIAL QUESTION
How do I classify objects?

Explore and Explain

 Tools   Watch

alike   different

Teacher Directions: Place some ▲ ■ ▼ along the grass on the page. Describe the blocks using the words *alike* and *different*. Place the pattern blocks that are the same on the tree with the word *alike*. Place the other pattern blocks on the tree with the word *different*. Trace and color the blocks.

Online Content at ⟋ **connectED.mcgraw-hill.com**   Chapter 9 • Lesson 1   539

See and Show

1 alike different

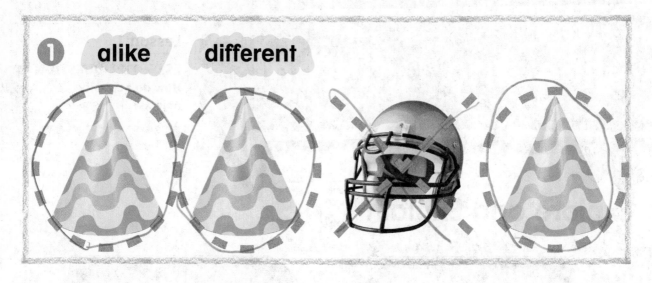

2

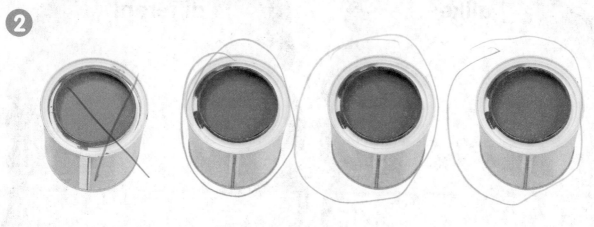

3

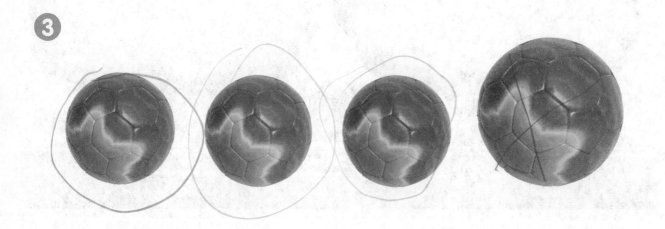

 Directions: 1. Look at the objects. Trace each circle to show the objects that are alike. Trace the X to show the object that is different. Tell why it does not belong. **2–3.** Look at the objects. Circle the objects that are alike. Draw an X on the one that is different. Tell why it does not belong.

Name

....

On My Own

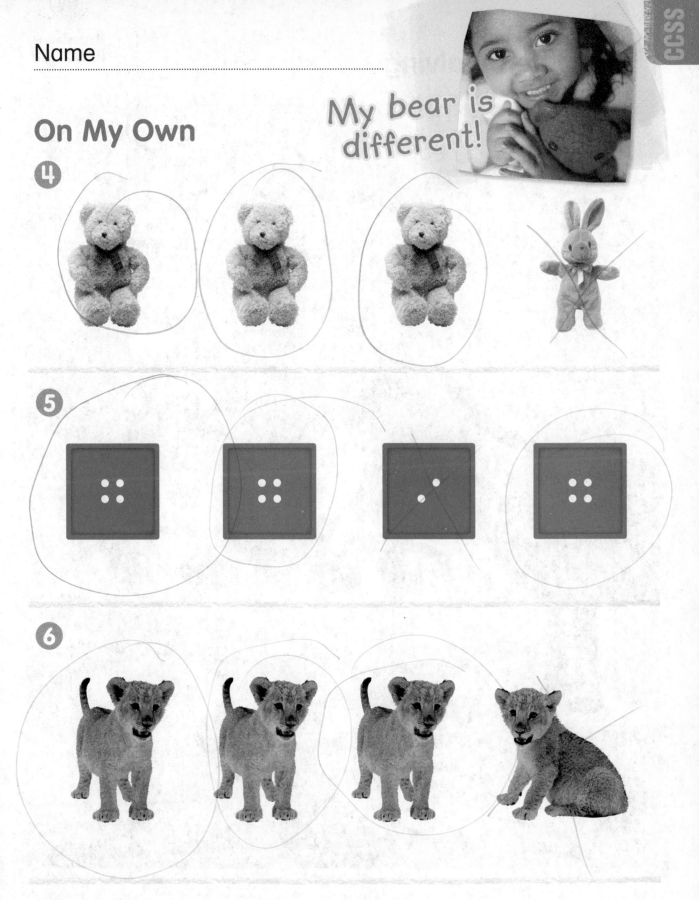

My bear is different!

④

⑤

⑥

 Directions: 4–6. Look at the objects. Circle the objects that are alike. Draw an X on the one that is different. Tell why it does not belong.

Directions: 7. Look at each group of objects near each animal. Circle the objects that are alike. Draw an X on each object that is different. Tell why it does not belong. Point to the shirt that is different. Draw the missing parts to make it the same as the other shirts.

Name _____

Measurement and Data
K.MD.3

CCSS

My Homework

Lesson 1
Alike and Different

Homework Helper eHelp   Need help? connectED.mcgraw-hill.com

1

2

3

Directions: 1–3. Look at the objects. Circle the objects that are alike. Draw an X on the one that is different. Tell why it does not belong.

Vocabulary Check

5 alike

6 different

Directions: 4. Look at the objects. Circle the objects that are alike. Draw an X on the one that is different. Tell why it does not belong. **5.** Look at the objects. Circle the objects that are alike. **6.** Look at the objects. Draw an X on the one that is different.

Math at Home Draw a picture of some items in each room of your home. Ask your child to tell how the items are alike or different.

Copyright © The McGraw-Hill Companies, Inc. (tl) Burke/Triolo Productions/Brand X Pictures/Getty Images; (tl) Burke/Triolo Productions/Brand X Pictures/Getty Images; (tc) Slede Preis/Getty Images; (cl) CMCD/Getty Images; (bl) Mark Steinmetz/The McGraw-Hill Companies, Inc.; (br) Photodisc/Getty Images

Name

Problem Solving
STRATEGY: Use Logical Reasoning

Lesson 2

ESSENTIAL QUESTION
How do I classify objects?

What does not belong?

Use Logical Reasoning

 Teacher Directions: Look at the objects below the picture. Color the objects that do not belong in the cold. Explain your answer.

What does not belong?

Use Logical Reasoning

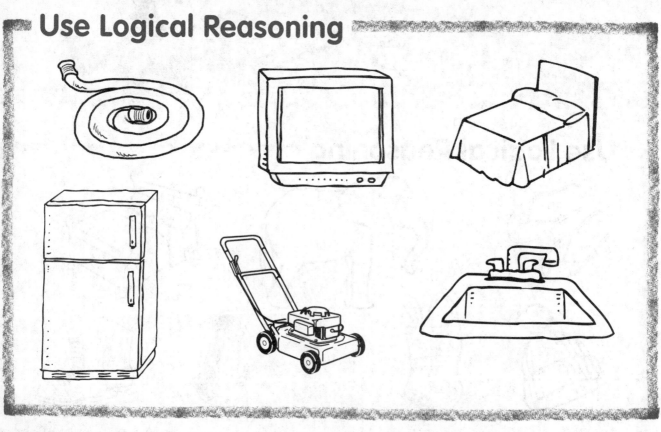

Directions: Look at the objects below the picture. Color the objects that do not belong in the yard. Explain your answer.

What does not belong?

Use Logical Reasoning

 Directions: Look at the objects below the picture. Color the objects that do not belong on the beach. Explain your answer.

What does not belong?

Use Logical Reasoning

 Directions: Look at the objects below the picture. Color the objects that are not classroom supplies. Explain your answer.

My Homework

What does not belong?

Use Logical Reasoning

Directions: Look at the objects below the picture. Color the objects that do not belong on a soccer field. Explain your answer.

What does not belong?

Use Logical Reasoning

 Directions: Look at the objects below the picture. Color the objects that do not belong in the park. Explain your answer.

Math at Home   Take advantage of problem-solving opportunities during daily routines such as putting away groceries. Ask what belongs in the cupboard or in the freezer.

Name _____

Sort by Size

Lesson 3

ESSENTIAL QUESTION
How do I classify objects?

Explore and Explain

Treasure hunt!

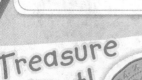

 Teacher Directions: Use six ▲ of the same color. Use two large buttons and four small buttons. Sort the buttons by size. Place the sorted buttons in the treasure box. Trace and color them to show how you sorted.

See and Show

1 sort   size

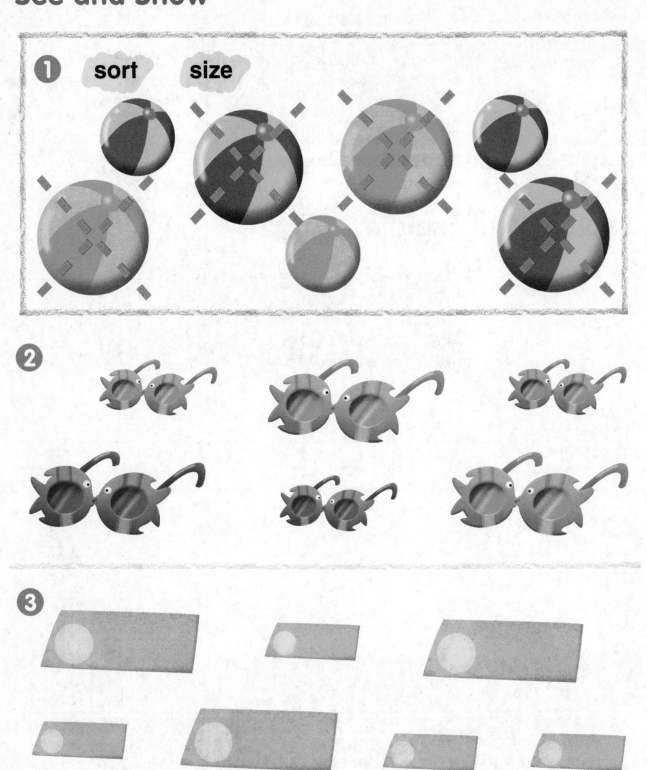

2

3

Directions: 1. The beach balls were sorted by size. Trace the Xs to show how the beach balls were sorted. **2–3.** Sort the objects by size. Draw Xs to show how you sorted.

Name

On My Own

4

5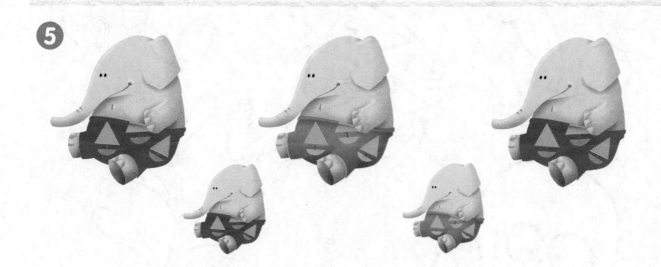

6

 Directions: 4–6. Sort the objects by size. Draw Xs to show how you sorted.

 Directions: 7. Look at the pairs of objects. Describe their sizes. Sort the objects by size. Color the large objects purple and the small objects yellow.

Measurement and Data
K.MD.3

CCSS

My Homework

Lesson 3

Sort by Size

Homework Helper

eHelp

Need help? connectED.mcgraw-hill.com

1

2

3

Directions: 1-3. Sort the objects by size. Draw Xs to show how you sorted. Tell how you sorted.

Vocabulary Check ![Vocab abc]

5 sort

6 size

Directions: 4. Sort the objects by size. Draw Xs to show how you sorted. Tell how you sorted. **5.** Sort the recycle bins by size. Draw an X on the large recycle bins. **6.** Draw an X on the objects that are the same size.

Math at Home Place kitchen towels and wash cloths in a group. Have your child sort the group by size.

Name ....

Check My Progress

Vocabulary Check

1 alike

2 sort

Concept Check

3

 Directions: 1. Circle the flowers that are alike. **2.** Sort the pillows by size. Draw Xs on the pillows to show how you sorted. **3.** Circle the objects that are alike. Draw an X on the one that is different.

4

5

6

 Directions: 4–5. Circle the objects that are alike. Draw an X on the one that is different. **6.** Sort the bugs by size. Draw Xs to show how you sorted.

Measurement and Data
K.MD.3

CCSS

Sort by Shape

Lesson 4

ESSENTIAL QUESTION
How do I classify objects?

Sorting is fun!

Explore and Explain

 Tools Watch

Teacher Directions: Use red ▮ of two different shapes. Place the attribute blocks in the red box. Sort the blocks by shape into groups. Place one group in the other box. Trace the shapes to show how you sorted.

Online Content at connectED.mcgraw-hill.com

See and Show

1 shape

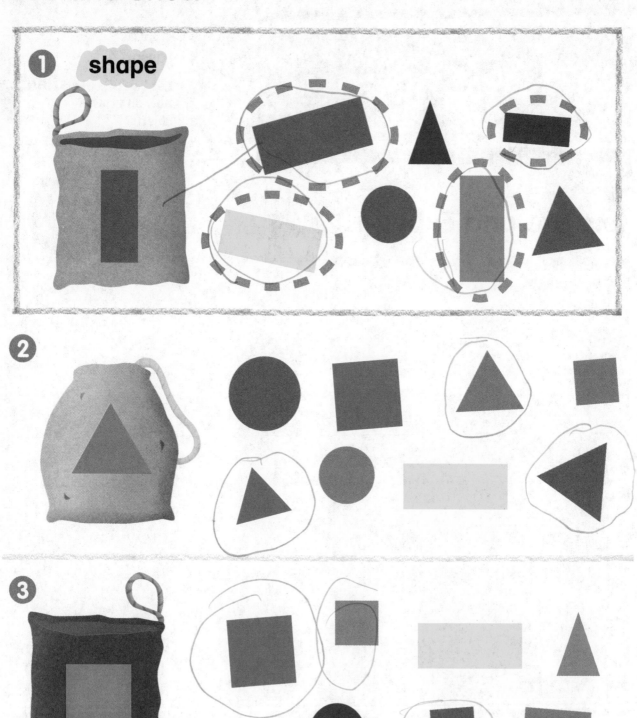

Directions: 1. Look at the shape on the bag. Sort the shapes in the group. Trace the circles around the shapes that belong in the bag. **2–3.** Look at the shape on the bag. Sort the shapes in the group. Draw a circle around the shapes that belong in the bag. Tell how you decided.

Name

....

On My Own

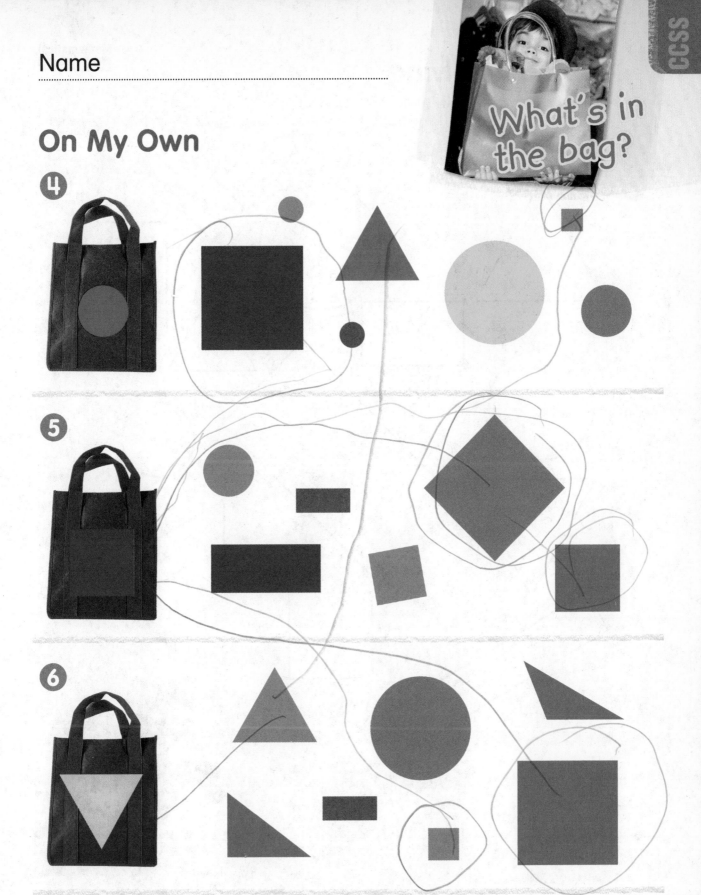

4

5

6

Directions: 4–6. Look at the shape on the bag. Sort the shapes in the group.
Draw a circle around the shapes that belong in the bag. Tell how you decided.

 **Problem Solving**

7

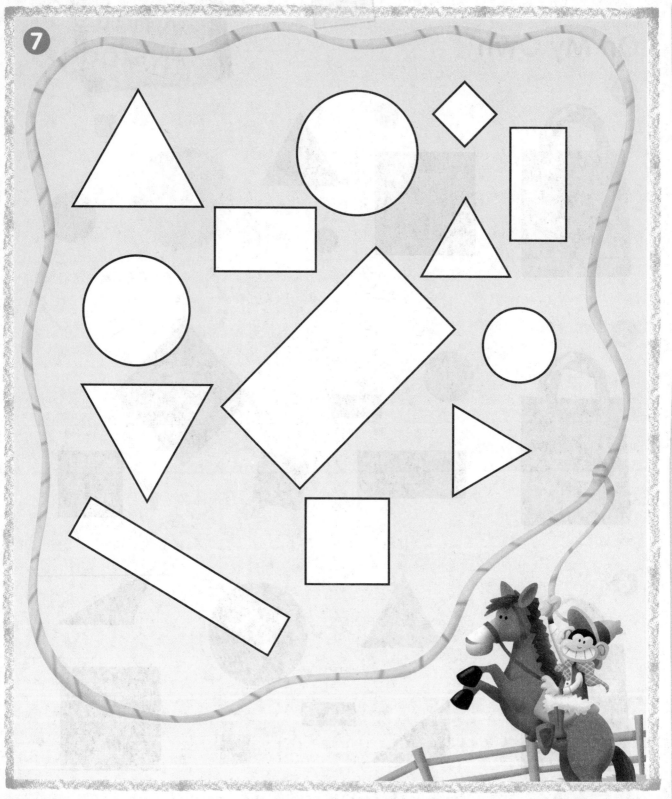

 Directions: 7. Sort the shapes. Color all ○ red, all ▭ orange, all □ green, and all △ blue.

Measurement and Data
K.MD.3

CCSS

My Homework

Homework Helper

eHelp

Need help? connectED.mcgraw-hill.com

1

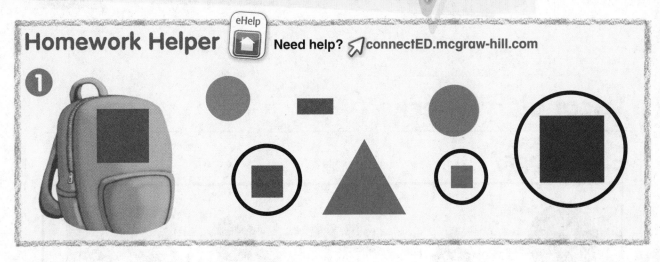

2

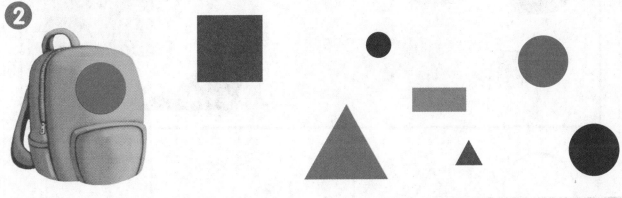

3

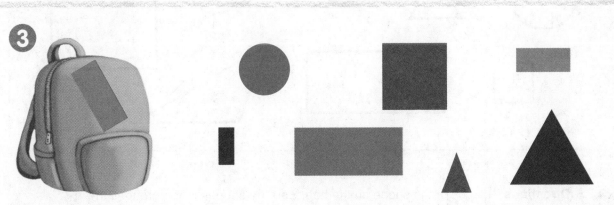

Directions: 1–3. Look at the shape on the bag. Sort the shapes in the group. Draw a circle around the shapes that belong in the bag.

4

Vocabulary Check

5 **shape**

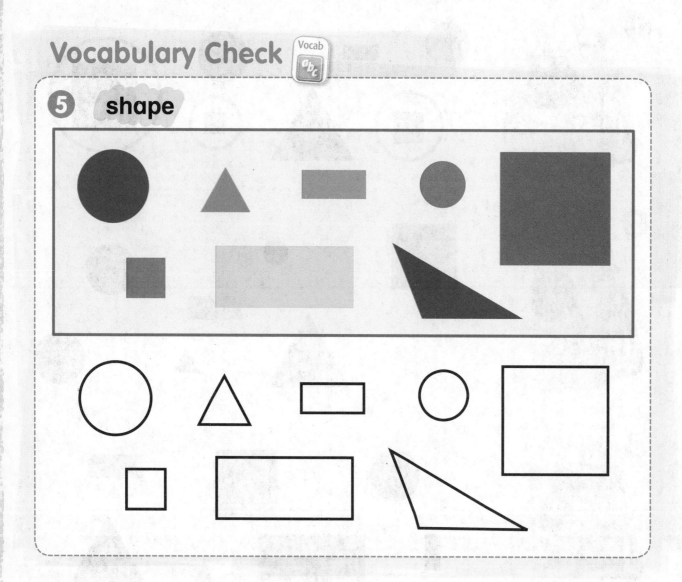

Directions: 4. Look at the shape on the bag. Sort the shapes in the group. Draw a circle around the shapes that belong in the bag. **5.** Look at the shapes in the red box. Color the shapes below to match.

Math at Home Mix round and square shaped cereal in a bowl. Have your child sort the cereal by shape into two different groups.

Name _____

Sort by Count

Lesson 5

ESSENTIAL QUESTION
How do I classify objects?

Explore and Explain

Tools

Teacher Directions: Use small attribute buttons. Place four ▦, three ●, and three ▲ randomly on the plate. Sort the buttons by shape into groups. Sort the groups by count. Tell how you sorted. Trace the buttons to show how you sorted.

See and Show

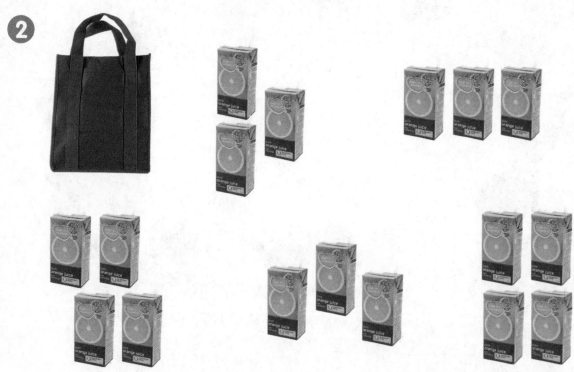

 Directions: 1. Count the bottles in each group. Trace the Xs to show how the bottles are sorted. **2.** Count the juice boxes in each group. Sort the juice boxes by count. Draw Xs to show how you sorted.

On My Own

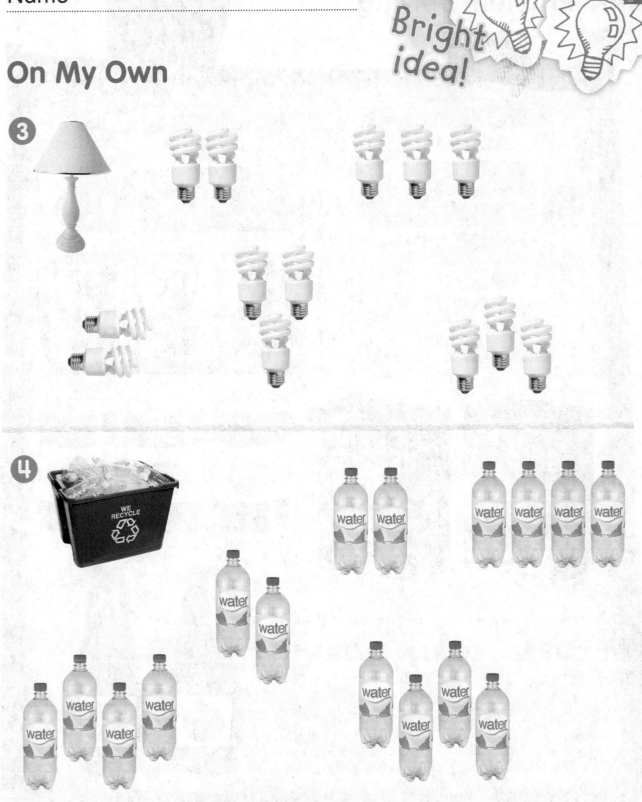

Directions: 3. Count the light bulbs in each group. Sort the light bulbs by count. Draw Xs to show how you sorted. **4.** Count the bottles in each group. Sort the bottles by count. Draw Xs to show how you sorted.

Problem Solving

5

Directions: 5. Count the objects in each group of food. Sort the objects by count. Draw Xs to show how you sorted. Sort by count again and circle the groups to show how you sorted.

Name

My Homework

Homework Helper

eHelp

Need help? connectED.mcgraw-hill.com

Directions: 1. Count the dogs. Sort the dogs by count. Draw Xs to show how you sorted. **2.** Count the fish. Sort the fish by count. Draw Xs to show how you sorted.

3

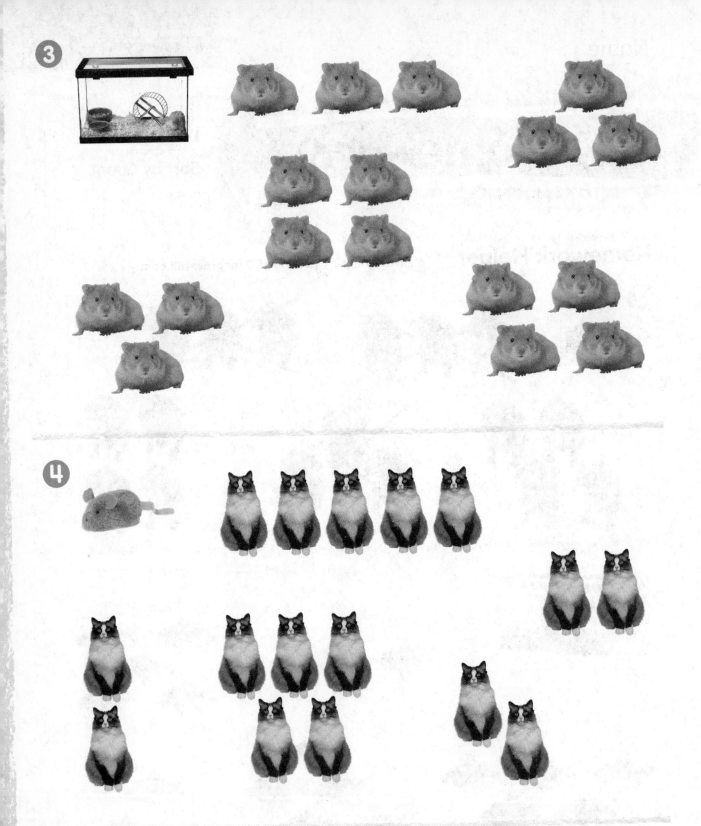

4

Directions: 3. Count the hamsters. Sort the hamsters by count. Draw Xs to show how you sorted. **4.** Count the cats. Sort the cats by count. Draw Xs to show how you sorted.

Math at Home Use buttons or another small object. Show a group of three, a group of four, a group of three, a group of four, and another group of four. Have your child sort the groups by count.

570   Chapter 9 • Lesson 5

My Review

Vocabulary Check

 Directions: 1. Look at the attribute buttons. Point to a shape. Draw an X on the buttons that are the same shape. **2.** Look at the blocks. Sort the blocks by size. Circle the blocks to show how you sorted.

Concept Check

1

2

3

 Directions: 1. Circle the objects that are alike. Draw an X on the one that is different. **2.** Sort the shapes. Color all △ blue, all □ red, and all ○ green. **3.** Sort the cans by size. Draw Xs on the cans to show how you sorted.

Name _____

Problem Solving

 Directions: 4. Look at the objects. Sort the objects by size. Draw lines from the objects to the recycle bins to show how you sorted.

shape   size

 Directions: Place attribute buttons on a tree. Sort the buttons by size or shape. Move the sorted buttons to the other tree. Trace the buttons on this tree to show how you sorted. Circle the word that tells how you sorted.

We See Animals in Action!

Watch a video!

Watch ▶

Geometry

K.G.1 Describe objects in the environment using names of shapes, and describe the relative positions of these objects using terms such as *above, below, beside, in front of, behind,* and *next to.*

Standards for Mathematical PRACTICE

1. Make sense of problems and persevere in solving them.
2. Reason abstractly and quantitatively.
3. Construct viable arguments and critique the reasoning of others.
4. Model with mathematics.
5. Use appropriate tools strategically.
6. Attend to precision.
7. Look for and make use of structure.
8. Look for and express regularity in repeated reasoning.

= focused on in this chapter

Name _____

Am I Ready?

Check ✓ ← Go online to take the Readiness Quiz

1

2

SCHOOL BUS

3

 Directions: 1. Draw eyes, a nose, and mouth on the face.
2. Color the bus wheels. **3.** Draw the missing chair legs.

My Math Words

 Vocab

Review Vocabulary

alike   different

 Directions: Trace each word. Discuss how the birds are alike and how they are different. Color the birds that are alike blue. Draw an X on the birds that are different. Explain your answer.

My Vocabulary Cards

Mathematical **PRACTICE**

above

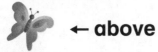

← above

behind

← behind

below

← below

beside

↑
beside

in front of

← in front of

next to

↑
next to

Teacher Directions:
Ideas for Use

- Ask students to arrange the cards to show the meaning of each word. For example, they might place the *above* card over of the *below* card.

- Have students sort the words by the number of letters in each word.

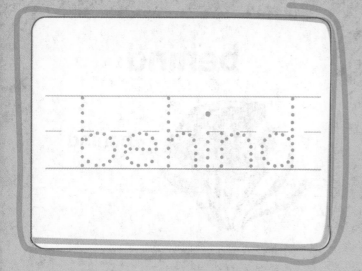

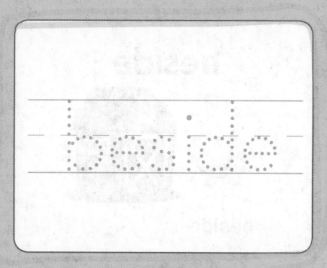

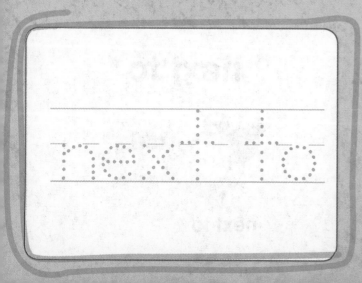

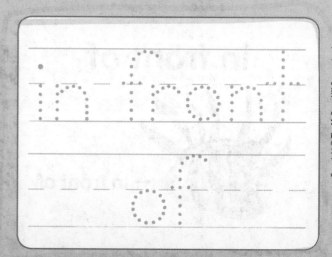

My Foldable

FOLDABLES Follow the steps on the back to make your Foldable.

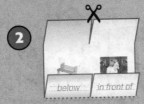

behind

above

in front of

below

Above and Below

Lesson 1

ESSENTIAL QUESTION
How do I identify positions?

Explore and Explain

 Teacher Directions: Place a on the page. Trace the cube. Use the words *above* or *over* and *below* or *under* to tell a partner where your cube is placed.

See and Show

1 above below

2

3

 Directions: 1. Describe the position of each monkey. Use *above* or *over* and *below* or *under*. Trace the circles around the monkeys that are above the branch. Trace the lines under the monkeys that are below the branch. **2.** Describe the position of each banana. Use *above* or *over* and *below* or *under*. Circle the bananas that are above a monkey. Underline the bananas that are below a monkey. **3.** Draw a banana below each monkey.

Name

On My Own

4

5

6

 Directions: 4. Circle the fish that is above a dolphin. Underline the fish that are below a dolphin. **5.** Circle the crabs that are above a snake. Underline the crab that is below a snake. **6.** Describe the position of each fish. Use the words *above* and *below* or *over* and *under*. Circle the fish that is above. Draw an X on the fish that is below.

Mathematical
PRACTICE

 Directions: 7. Draw a cloud above or over the airplane. Draw a rabbit below or under the grass. Draw a window above or over the door on the house. Circle the object that is above the house and below the sun.

My Homework

Homework Helper

eHelp

Need help? connectED.mcgraw-hill.com

1

2

3

Directions: 1. Circle the bees that are above a flower. Underline the bee that is below a flower. **2.** Circle the dragonflies that are above a branch. Underline the dragonfly that is below a branch. **3.** Circle the butterflies that are above the net. Underline the butterfly that is below the net.

④

Vocabulary Check

⑤ above

⑥ below

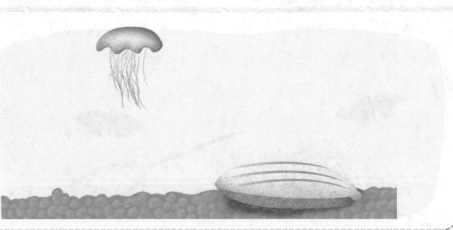

Directions: 4. Describe the position of each bird. Use *above* and *below*. Circle the bird that is above. Draw an X on the bird that is below. **5.** Draw a fish above the shark. **6.** Draw a seashell below the jellyfish.

Math at Home Give your child directions using the words *above* and *below*. For example, stack these books above the games on the shelf.

Geometry
K.G.1

CCSS

In Front of and Behind

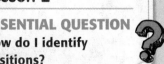

Lesson 2

ESSENTIAL QUESTION
How do I identify positions?

Explore and Explain

Tools Watch

 Teacher Directions: Draw a box around the object that is in front of the sand castle. Draw an X on each animal that is on the beach in front of the ladies. Draw an object behind the dolphin.

Online Content at connectED.mcgraw-hill.com

See and Show

①

②

③

④

 Directions: 1. Trace the box around the object that is in front of the school. **2.** Trace the X on the object that is behind the plastic tube. **3.** Draw an X on each student in front of the bus. **4.** Draw a box around the raccoon that is in front of the slide.

Name

_____

On My Own

5

6

7

8

 Directions: 5. Draw a box around the animal that is behind the alligator. **6.** Draw an X on the animal that is in front of the cat. **7.** Draw a box around the animal that is behind the hippo. **8.** Draw an object. Draw another object behind your object.

Problem Solving

Directions: 9. Draw an X on the girl that is in front of the house. Draw a box around the flower pot that is in front of the house. Draw a circle around the flower pot that is behind another flower pot. Circle the bear that is behind the raccoon.

Geometry
K.G.1

CCSS

My Homework

Lesson 2

In Front of and Behind

Homework Helper
eHelp

Need help? connectED.mcgraw-hill.com

1

2

3

Directions: 1. Draw a box around the fish that is behind the treasure chest. **2.** Draw a box around the mouse that is behind the snake. **3.** Draw a box around the animal that is in front of the barn.

④

Vocabulary Check

⑤ in front of

⑥ behind

Directions: 4. Draw an X on the animal that is behind the fence. **5.** Draw an apple in front of the pig. **6.** Draw a flower behind the goat.

Math at Home Have your child look in the cupboard or refrigerator. Choose an item that has an item in front of it and behind it. Have your child tell which objects are in front of and behind. Choose another object and repeat.

Name

Check My Progress

Vocabulary Check

1 **in front of**

2 **above**

Concept Check

3

Directions: **1.** Draw a ball in front of the bike. **2.** Draw a balloon above the boy.
3. Draw X on the animal that is behind the bird.

Directions: 4. Circle the balloon that is above or over the tree. **5.** Draw a box around the bird that is below or under the red bird. **6.** Draw an X on the butterfly above the flower. **7.** Draw a box around the bug that is below or under the other bug.

Name

Next to and Beside

Lesson 3

ESSENTIAL QUESTION
How do I identify positions?

Explore and Explain

 Teacher Directions: Circle the animal that is next to the tree. Draw an X on the animal that is beside the rock. Draw a frog next to the small elephant.

Online Content at connectED.mcgraw-hill.com

Chapter 10 • Lesson 3

See and Show

1 next to   beside

2

Directions: 1. Trace the circle around the pig that is next to the orange house. Trace the X on the house that is next to the pig with the yellow shirt. **2.** Draw an X on the animal that is next to the brick house. Draw a box around the pig that is beside the pig in the red shirt. Circle the animal that is next to the blue bird. Draw a flower beside the caterpillar.

598   Chapter 10 • Lesson 3

Name

....

On My Own

3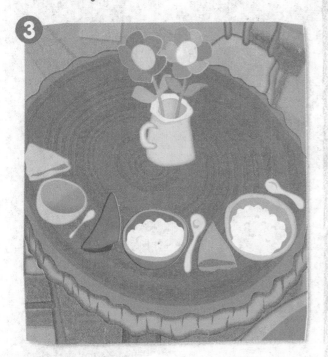

4

5

6

 Directions: 3. Draw an X on the spoon that is next to the orange bowl. **4.** Draw an X on the chairs that are beside the green chair. **5.** Draw an X on the bed that is next to the bed with the flower blanket. **6.** Draw three different colored flowers in a row. Ask a partner to tell you which flower is next to or beside one of the flowers.

Online Content at **connectED.mcgraw-hill.com**

 Directions: 7. Point to the part of the caterpillar's body that is next to the color pink. Color it orange. Point to the part of the caterpillar's body that is next to the head. Color it blue. Color all other parts any color. Tell a classmate which colors you chose, and tell the colors that are beside them.

Name

My Homework

Homework Helper

Need help? connectED.mcgraw-hill.com

1

2

3

Directions: 1. Draw a box around the animal that is next to the fox. Draw an X on the object that is beside the rabbit. **2.** Draw an X on the animal that is beside the lizard. Draw a box around the object that is next to the turtle. **3.** Draw an X on the object that is next to the lizard. Draw a box around the animal that is next to the rabbit in the hole.

4

Vocabulary Check

5 **next to**

6 **beside**

 Directions: 4. Draw an X on the animal next to the cactus. Draw a box around the animal that is beside the armadillo. **5.** Draw a bone next to the dog. **6.** Draw a ball beside the cat.

Math at Home At home, give your child directions using the words *next to* and *beside*. For example, put your shoes on the floor next to the other shoes in the closet.

Problem Solving
STRATEGY: Act It Out

Where do I put it?

Act It Out

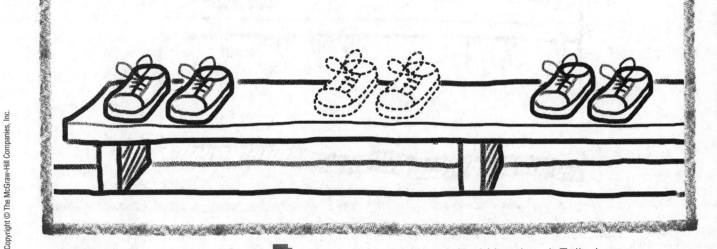

 Teacher Directions: Use ■ to show where the shoes should be placed. Tell where they belong. Trace the shoes where they belong.

Copyright © The McGraw-Hill Companies, Inc.

Where do I put it?

Act It Out

 Directions: Use a connecting cube to show where the cat should be placed. Draw the cat where it belongs. Tell where the cat belongs.

Name ....

Where do I put it?

Act It Out

 Directions: Use a connecting cube to show where the lunch box should be placed. Draw the lunch box where it belongs. Tell where the lunch box belongs.

Where do I put it?

Act It Out

 Directions: Use connecting cubes to show where the flowers should be placed. Draw the flowers where they belong. Tell where the flowers belong.

Name

My Homework

Where do I put it?

Act It Out

Directions: Use dry cereal to show where the box of cereal should be placed. Trace the box to show where it belongs. Tell where the cereal belongs.

Where do I put it?

Act It Out

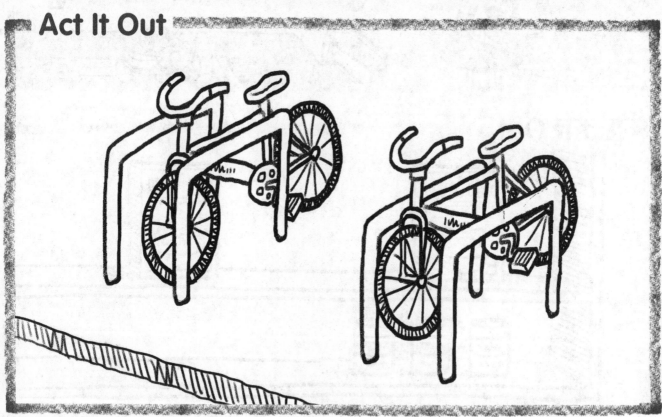

 Directions: Use dry cereal to show where the bike should be placed. Draw the bike where it belongs. Tell where the bike belongs.

Math at Home Take advantage of problem-solving opportunities during daily routines such as riding in the car, bedtime, doing laundry, and so on.

Name

....

My Review

Vocabulary Check

beside

above

below

behind

next to

in front of

 Directions: 1. Draw another bird flying above the trees. **2.** Circle the animal behind the tree. **3.** Draw a coconut beside a monkey. **4.** Draw a monkey next to one of the lion cubs. **5.** Circle the monkey below the branch. **6.** Draw an X on the lions in front of the tree.

Concept Check

①

②

③

④

Directions: 1. Draw an X on the bubble that is above or over the bottle. **2.** Draw a box around the flower that is below or under the green flower. **3.** Draw a crayon next to the box. **4.** Draw a dog bone beside the bowl.

Name

Problem Solving

Directions: 5. Circle the object above the dog. Draw an X on the animal that is next to the black cat. Draw a bird below the sun. Underline the object that is on the blanket next to the girl with the red book.

Reflect

 Directions: Draw an X on the butterfly that is above the flower pot. Draw a box around the animal that is in front of the baby bear. Draw a circle around the bees that are below the bee hive. Circle the window that is beside the door.

Chapter

11

Two-Dimensional Shapes

Let's Discover Shapes!

Watch a video!

Watch

My Common Core State Standards

CCSS

Geometry

K.G.1 Describe objects in the environment using the names of shapes, and describe the relative position of these objects using terms such as above, below, in front of, behind, and next to.

K.G.2 Correctly name shapes regardless of their orientations or overall size.

K.G.3 Identify shapes as two-dimensional (lying in a plane, "flat") or three-dimensional ("solid").

K.G.4 Analyze and compare two- and three-dimensional shapes, in different sizes and orientations, using informal language to describe their similarities, differences, parts (e.g., number of sides and vertices/"corners") and other attributes (e.g., having sides of equal length).

K.G.5 Model shapes in the world by building shapes from components (e.g., sticks and clay balls) and drawing shapes.

K.G.6 Compose simple shapes to form larger shapes.

Standards for Mathematical PRACTICE ⬇

1. Make sense of problems and persevere in solving them.
2. Reason abstractly and quantitatively.
3. Construct viable arguments and critique the reasoning of others.
4. Model with mathematics.
5. Use appropriate tools strategically.
6. Attend to precision.
7. Look for and make use of structure.
8. Look for and express regularity in repeated reasoning.

⬤ = focused on in this chapter

Name _____

Am I Ready?

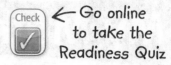  Check ← Go online to take the Readiness Quiz

1

2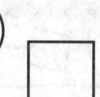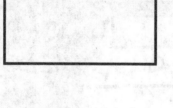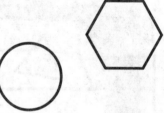

3

4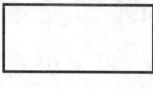

 Directions: 1–2. Circle the objects that are the same shape. **3–4.** Look at the shapes. Find the shapes in the row that are the same. Color them blue.

Name

My Math Words

Vocab abc

Review Vocabulary

each   size

Directions: Trace each word. Use the words to describe the objects in each box.

circle

hexagon

rectangle

round

side

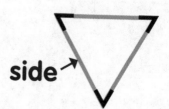

side

straight

Teacher Directions:
Ideas for Use

- Group three words that belong together. Add a word that does not belong in the group. Ask another student to name the word that does not belong.

- Draw a line on each card every time you read or write the word in this chapter. Try to use at least 10 lines for each card.

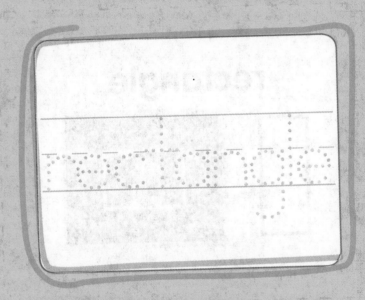

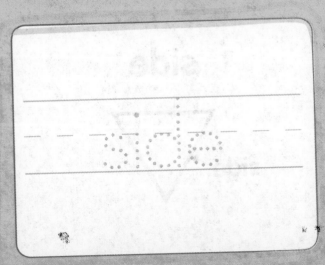

Vocab

square

triangle

 YIELD

vertex

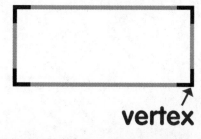

vertex

Teacher Directions:
More Ideas for Use

- Have students sort the words by the number of letters in each word.

- Have students write a letter on the front and then draw a picture of an object that starts with that letter on the back.

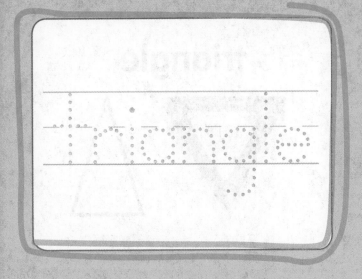

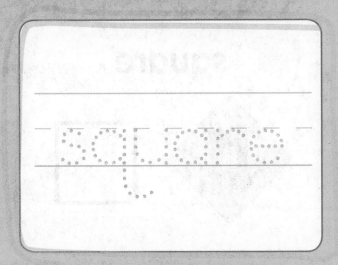

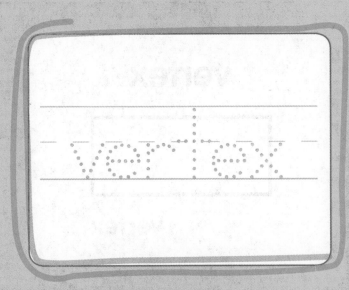

My Folding Book

FOLDABLES® Follow the steps on the back to make your Foldable.

square

triangle

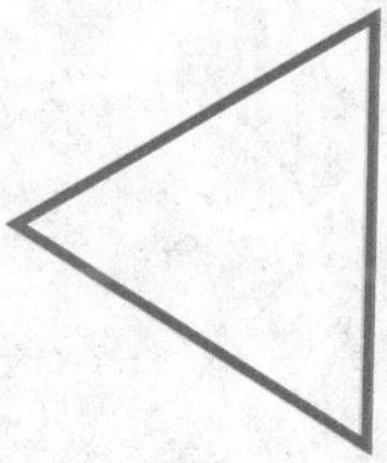

rectangle

circle

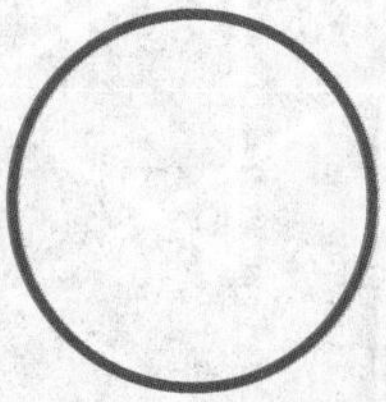

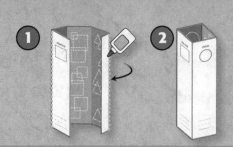

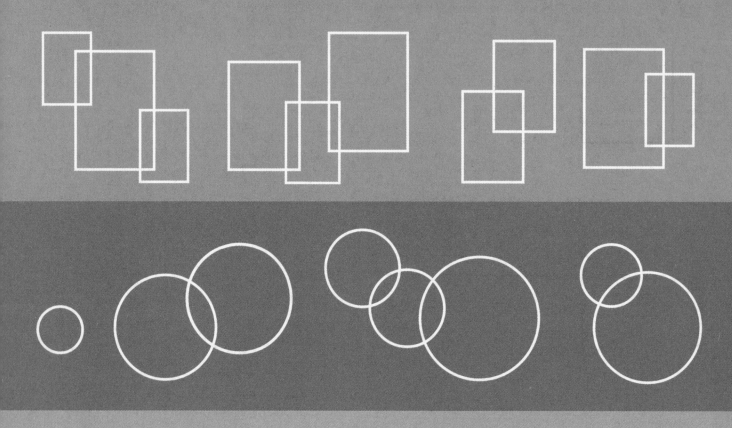

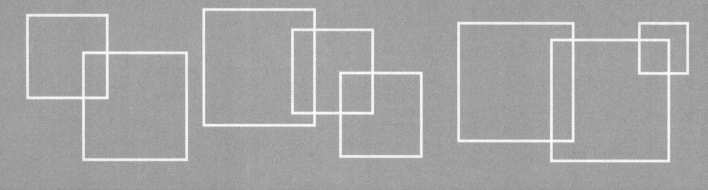

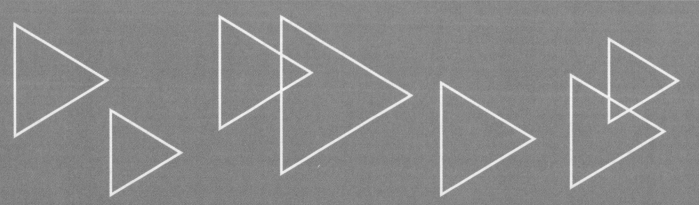

Name _____

Squares and Rectangles

Lesson 1

ESSENTIAL QUESTION
How can I compare shapes?

Explore and Explain

Tools Watch

 Teacher Directions: Use [ ] [ ] to make a picture. Trace the shapes. Identify the shapes by coloring the squares blue and rectangles orange. Describe the shapes using the words *longer sides* and *same length sides*. How many sides? How many vertices?

See and Show

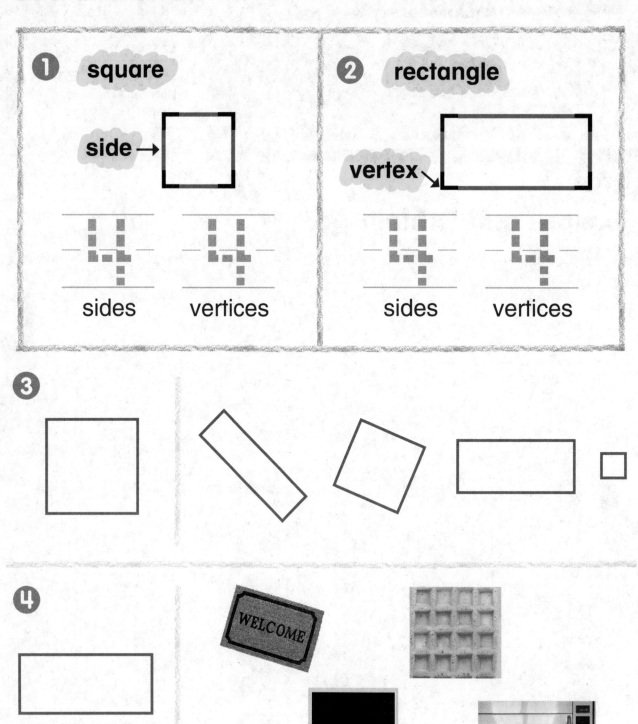

① **square**

side →

4 **sides**   4 **vertices**

② **rectangle**

vertex →

4 **sides**   4 **vertices**

③

④

WELCOME

Directions: 1–2. Name the shape. Trace the number of sides and vertces. **3.** Name the shape. Describe it. Compare the shape to each shape in the group. Circle the matching shapes. **4.** Name the shape. Describe it. Compare the shape to each object in the group. Circle the objects that are the same shape.

....

On My Own

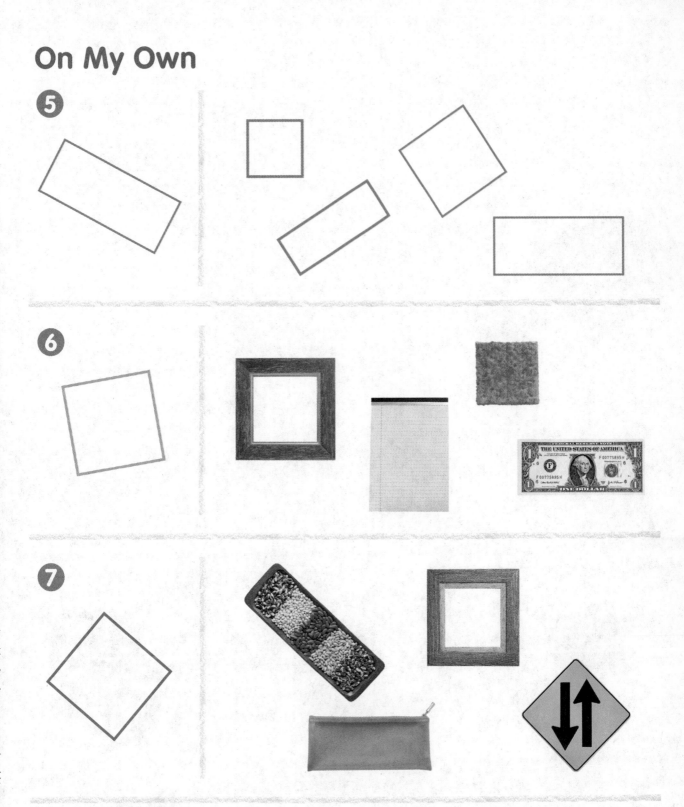

5

6

7

Directions: 5. Name the shape. Describe it. Compare the shape to each shape in the group. Circle the matching shapes. **6–7.** Name the shape. Describe it. Compare the shape to each object in the group. Circle the objects that are the same shape.

Online Content at connectED.mcgraw-hill.com

Real World **Problem Solving**

8

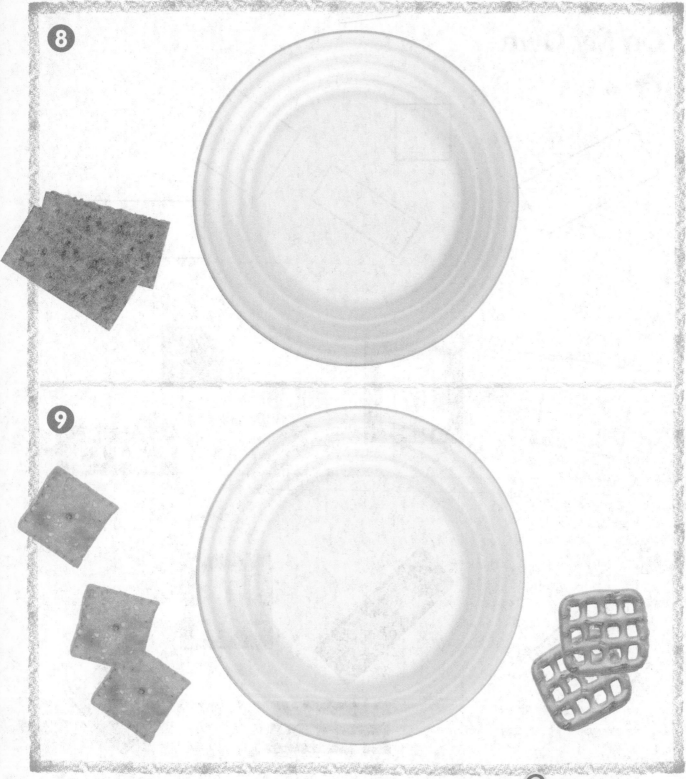

9

Directions: 8. Draw rectangle shaped foods on the plate.
9. Draw square shaped foods on the plate.

Can you think of more food shapes?

Name

My Homework

Lesson 1

Squares and
Rectangles

Homework Helper

eHelp

Need help? connectED.mcgraw-hill.com

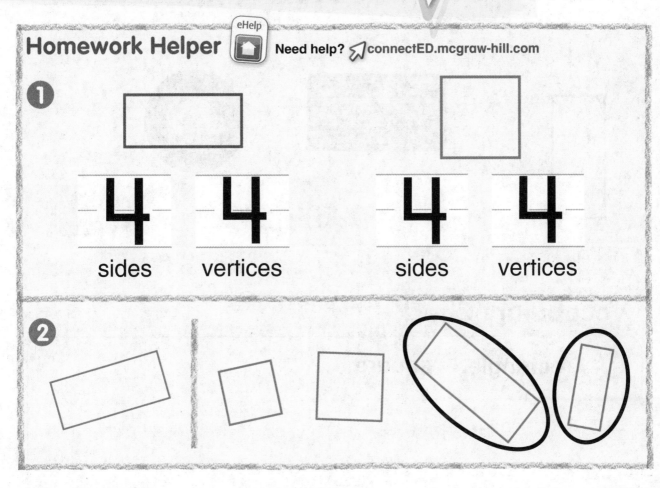

1

4 4 4 4

sides   vertices   sides   vertices

2

3

 Directions: 1. Name each shape. Write the number of sides and vertices. **2–3.** Name the shape. Describe it. Compare the shape to each shape in the group. Circle the matching shapes.

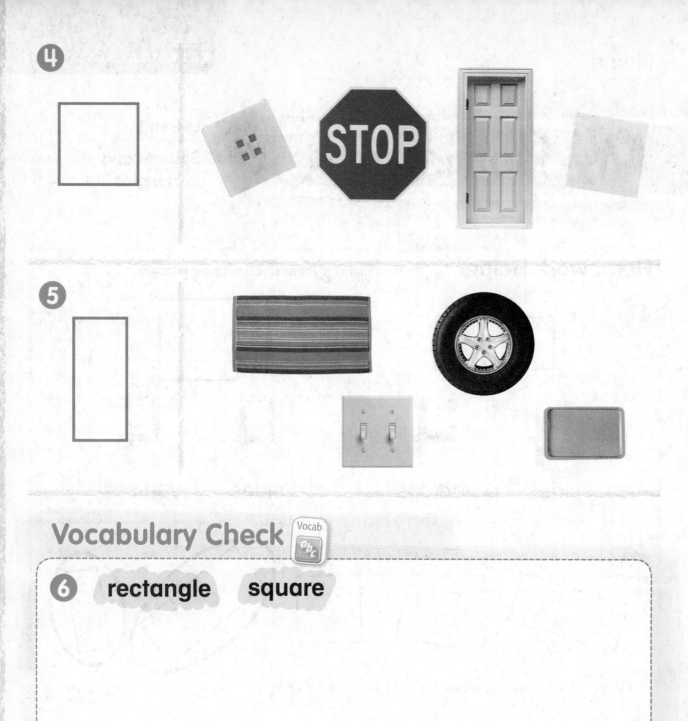

4

5

Vocabulary Check

6   rectangle   square

Directions: 4–5. Name the shape. Describe it. Compare the shape to each object in the group. Circle the objects that are the same shape. **6.** Draw a rectangle. Draw a square. Circle the shape that has the same length sides.

Math at Home   Look at flat objects in your home such as table tops, windows, cupboard doors, computer, and stair steps. Ask your child to identify the shape of each.

Geometry
K.G.2, K.G.3, K.G.4, K.G.5

CCSS

Circles and Triangles

Explore and Explain

Lesson 2

ESSENTIAL QUESTION
How can I compare shapes?

Teacher Directions: Use  to make a picture. Trace the shapes. Identify the shapes by coloring the circles red and the triangles purple. Describe the shapes using the words *round* and *straight*. Count how many sides and vertices.

See and Show

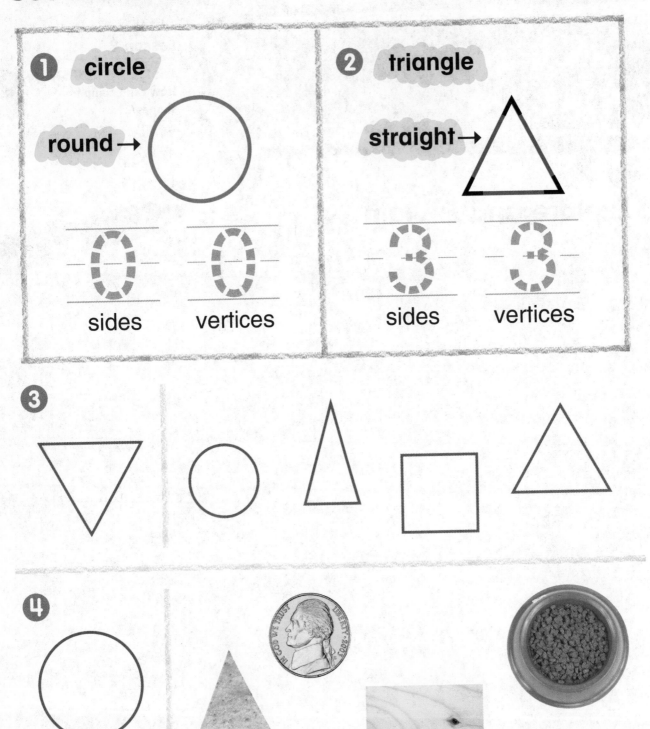

1 circle

round →

_____ sides _____ vertices

2 triangle

straight →

_____ sides _____ vertices

3

4

Directions: 1–2. Name the shape. Trace the number of sides and vertices. **3.** Name the shape. Describe it. Compare the shape to each shape in the group. Circle the matching shapes. **4.** Name the shape. Describe it. Compare the shape to each object in the group. Circle the objects that are the same shape.

Name

On My Own

Hi there!

5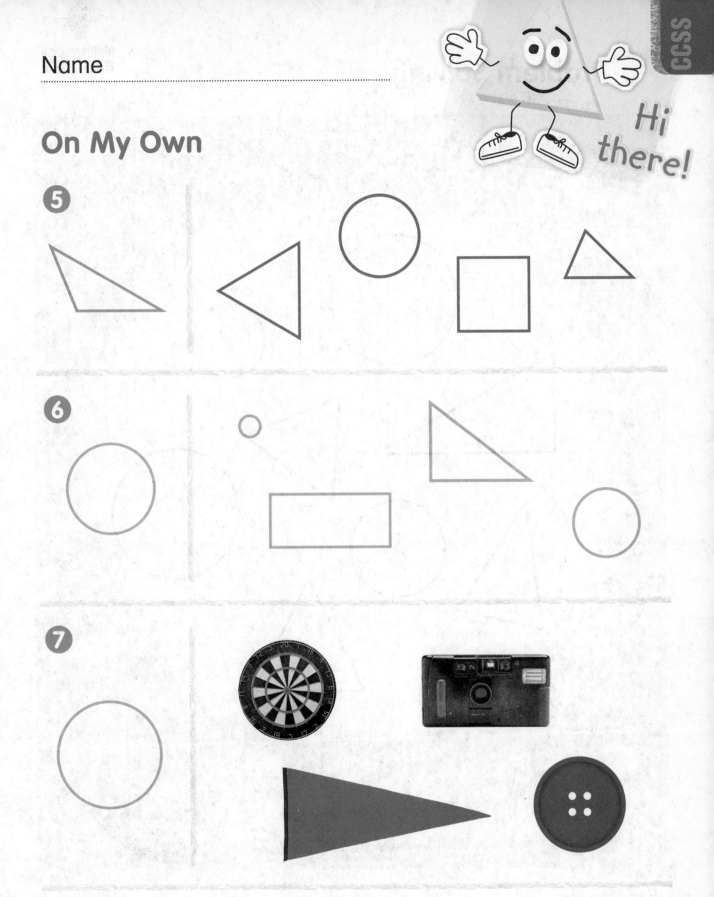

6

7

 Directions: 5–6. Name the shape. Describe it. Compare the shape to each shape in the group. Circle the matching shapes. **7.** Name the shape. Describe it. Compare the shape to each object in the group. Circle the objects that are the same shape.

Online Content at ↗ **connectED.mcgraw-hill.com**   Chapter 11 • Lesson 2   631

Problem Solving

8

Directions: 8. Identify the circles and triangles on the fruit pizza by coloring the circles green and the triangles orange.

Name

....

My Homework

Homework Helper

eHelp

Need help? connectED.mcgraw-hill.com

1

0	**0**	**3**	**3**
sides	vertices	sides	vertices

2

3

 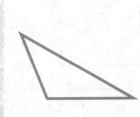 

 Directions: 1. Name each shape. Write the number of sides and vertices. **2–3.** Name the shape. Describe it. Compare the shape to each shape in the group. Circle the matching shapes.

④

⑤

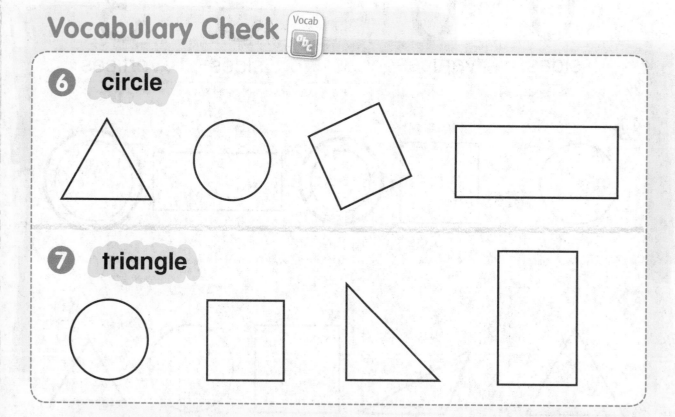

Vocabulary Check

⑥ circle

⑦ triangle

Directions: 4–5. Name the shape. Describe it. Compare the shape to each object in the group. Circle the objects that are the same shape. **6.** Color the circle red. **7.** Color the triangle blue.

Math at Home Gather a group of objects of various shapes and sizes. Include triangles and circles. Ask your child to sort the triangles and circles.

Name

Squares, Rectangles, Triangles, and Circles

Lesson 3

ESSENTIAL QUESTION
How can I compare shapes?

Explore and Explain

Tools Watch

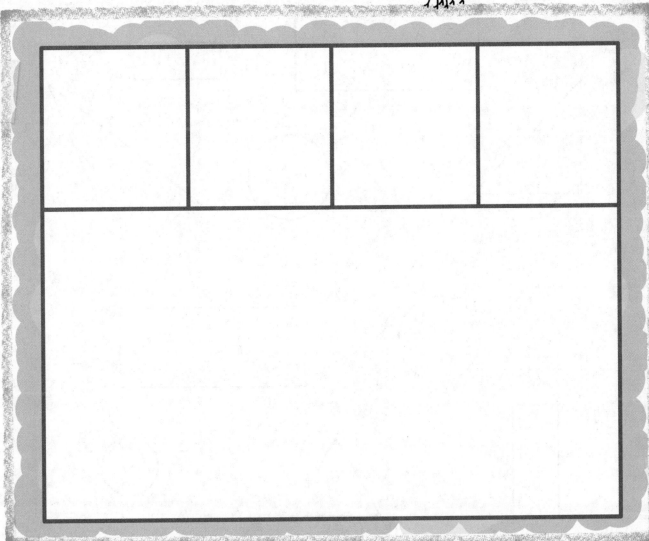

 Teacher Directions: Place   at the bottom of the mat. Sort the attribute blocks. Place one of each shape in the boxes at the top. Trace the shapes. Name the shapes. Count the sides and vertices of each shape. Compare the shapes.

See and Show

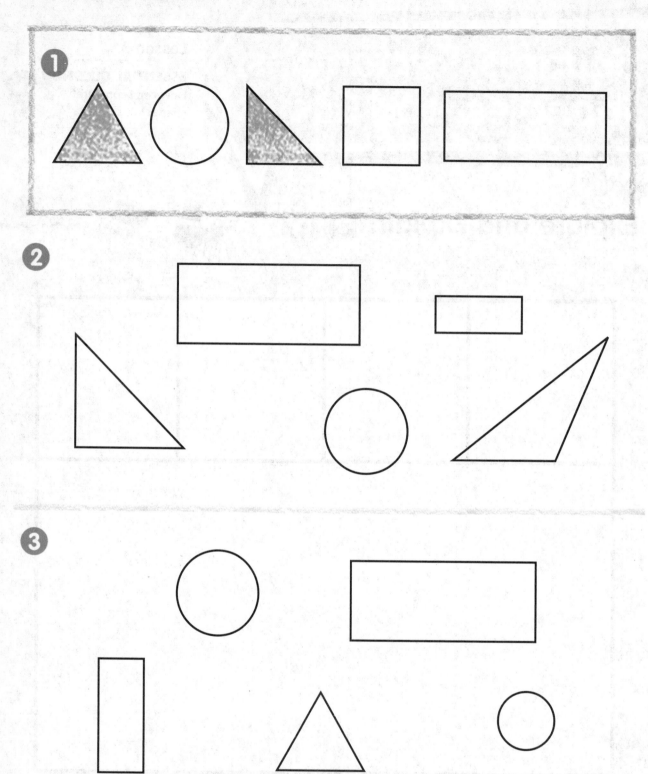

1

2

3

 Directions: 1. Color the shapes that have three sides and three vertices blue. **2.** Color the shapes that have four sides and four vertices red. **3.** Color the shapes that have zero sides and zero vertices green.

Name _____

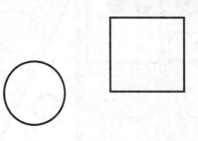

On My Own

4

5

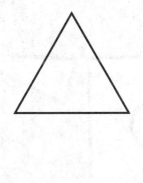
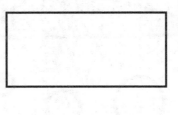
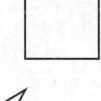

6

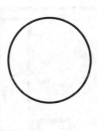
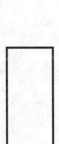
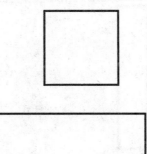
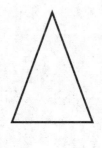

 Directions: 4. Color the shapes that are round purple.
5. Color the shapes that have three sides and three vertices red.
6. Color the shapes that have four sides and four vertices orange.

7

Great work!

Directions: 7. Color the square(s) red, the triangle(s) yellow, the rectangle(s) green, and the circle(s) blue.

Name

My Homework

Lesson 3

Squares, Rectangles, Triangles, and Circles

Homework Helper

eHelp

Need help? connectED.mcgraw-hill.com

1

square   circle   triangle   rectangle

2

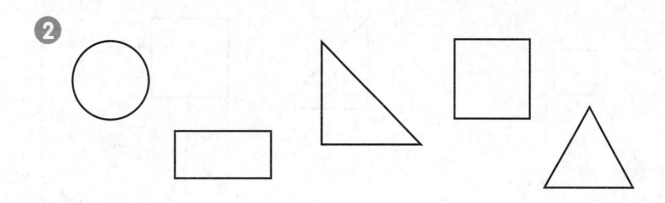

3

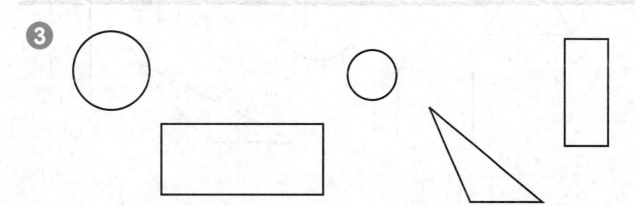

Directions: 1. Color the shapes that have four sides and four vertices blue.
2. Color the shapes that have three sides and three vertices red.
3. Color the shapes that have zero sides and zero vertices green.

Chapter 11 • Lesson 3   639

④

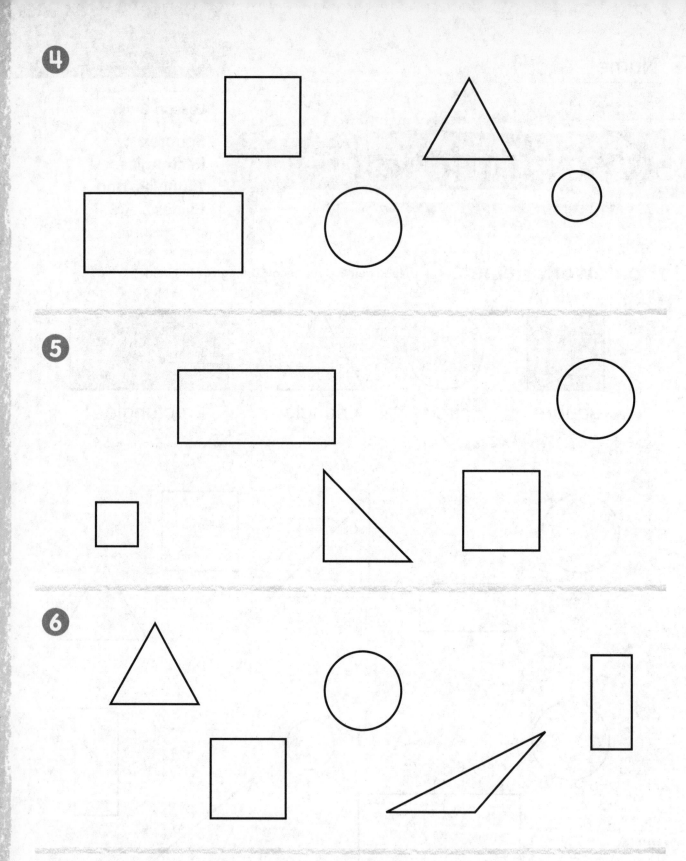

⑤

⑥

Directions: 4. Color the shapes that are round purple.
5. Color the shapes that have four sides and four vertices yellow.
6. Color the shapes that have three sides and three vertices orange.

Math at Home While driving in the car, have your child look for shapes. For example, street signs, wheels, etc. Have him or her identify each shape seen.

640   Chapter 11 • Lesson 3

Name

Hexagons

Lesson 4

ESSENTIAL QUESTION
How can I compare shapes?

Explore and Explain

 Tools Watch

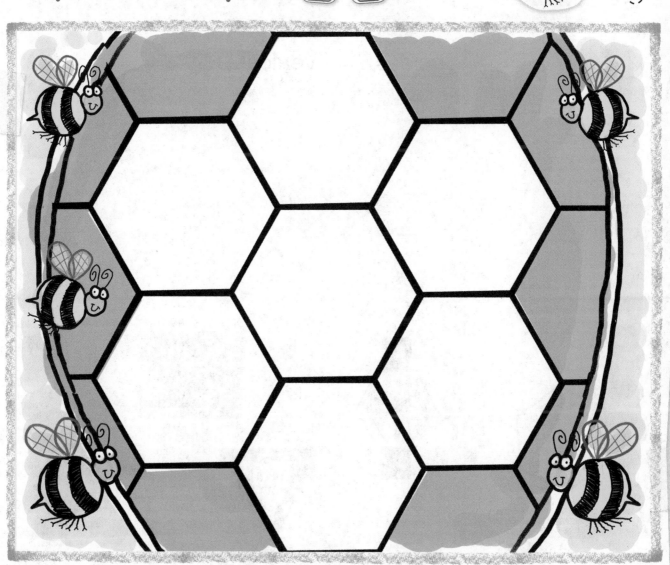

 Teacher Directions: Use and fill in the beehive. Color each hexagon a different color. Count the sides and vertices. Describe the shape using the words *sides* and *vertices*.

See and Show

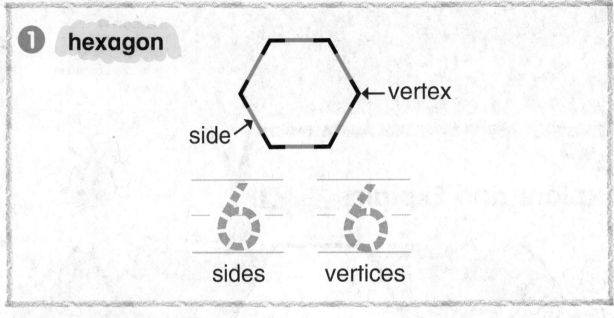

1 hexagon

vertex

side

_6_ _6_

sides   vertices

2

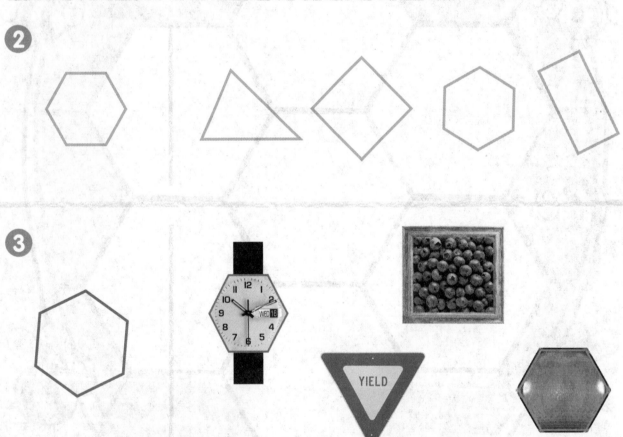

3

YIELD

Directions: 1. Name the shape. Trace the number of sides and vertices. **2.** Name the shape. Describe it. Compare the shape to each shape in the group. Circle the matching shape. **3.** Name the shape. Describe it. Compare the shape to each object in the group. Circle the objects that are the same shape.

Name

On My Own

 4

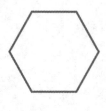    

 5

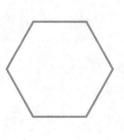 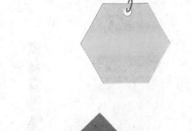  

 6

 Directions: 4. Name the shape. Describe it. Compare the shape to each shape in the group. Circle the matching shape. **5–6.** Name the shape. Describe it. Compare the shape to each object in the group. Circle the object that is the same shape.

7

Directions: 7. Use hexagon and triangle pattern blocks to create flowers. Trace the blocks. Color the hexagons orange. Color the triangles blue. Name the shapes you used.

Name
....

My Homework

Lesson 4

Hexagons

Homework Helper eHelp   Need help? connectED.mcgraw-hill.com

1

← vertex

side

6 sides   **6** vertices

2

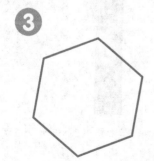 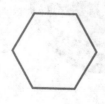    

3

Directions: 1. Name the shape. Write the number of sides and vertices. **2–3.** Name the shape. Describe it. Compare the shape to each shape in the group. Circle the matching shape.

4 **5** **6**

Directions: 4. Name the shape. Describe it. Compare the shape to each shape in the group. Circle the matching shape. **5–6.** Name the shape. Describe it. Compare the shape to each object in the group. Circle the object that is the same shape.

Math at Home While in the grocery store, have your child look to see if he or she can see shapes. Have your child look for hexagons.

Name

Check My Progress

Vocabulary Check

1 circle	**2** triangle

Concept Check

3

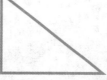

4

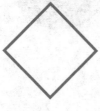   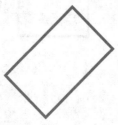

 Directions: 1. Draw circles and color them blue. **2.** Draw triangles and color them orange. **3–4.** Name the shape. Describe it. Compare the shape to each shape in the group. Circle the matching shapes.

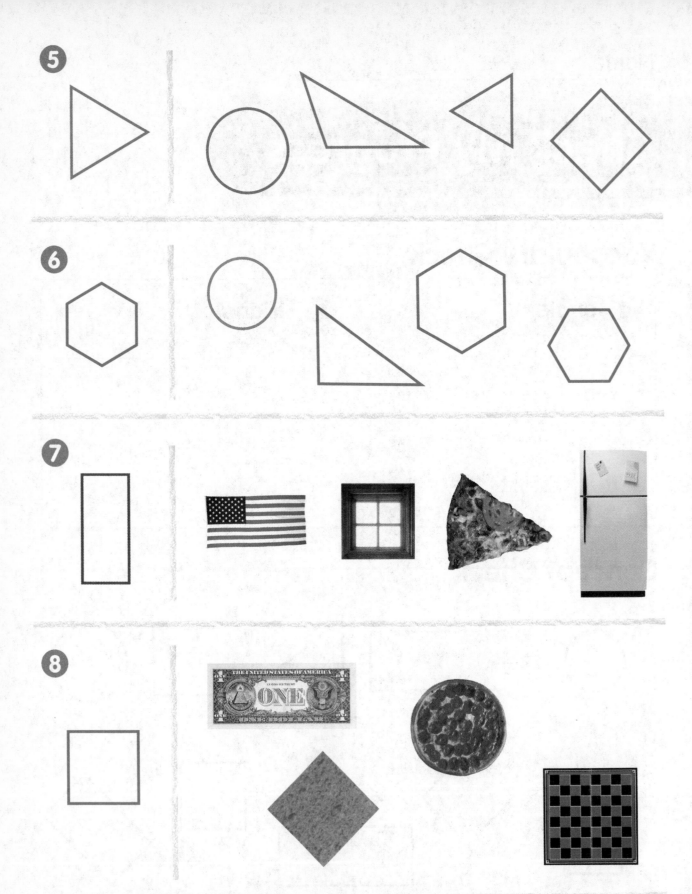

 Directions: 5–6. Name the shape. Describe it. Compare the shape to each shape in the group. Circle the matching shapes. **7–8.** Name the shape. Describe it. Compare the shape to each object in the group. Circle the objects that are the same shape.

Name

Shapes and Patterns

Lesson 5

ESSENTIAL QUESTION
How can I compare shapes?

Patterns are fun!

Explore and Explain

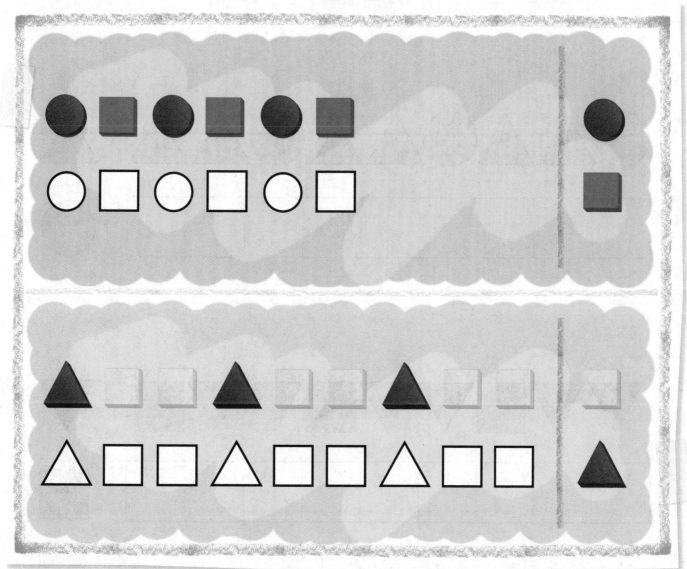

Teacher Directions: Use pattern blocks to copy the pattern. Color the shapes to match the pattern. Circle the shape that could come next.

See and Show

1

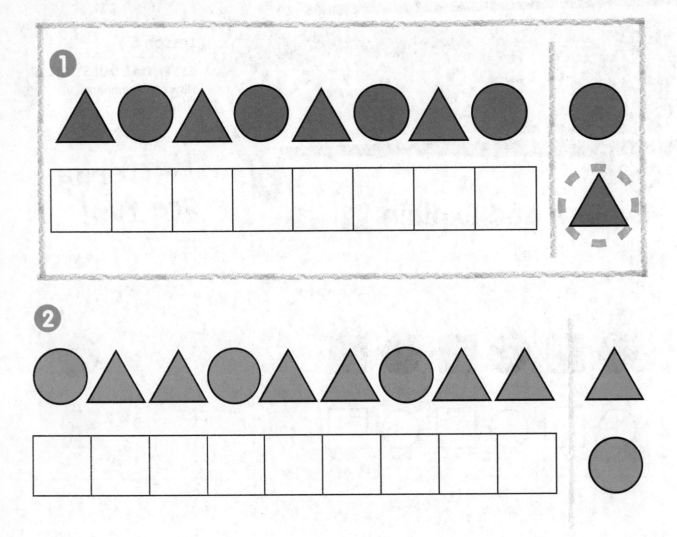

2

3

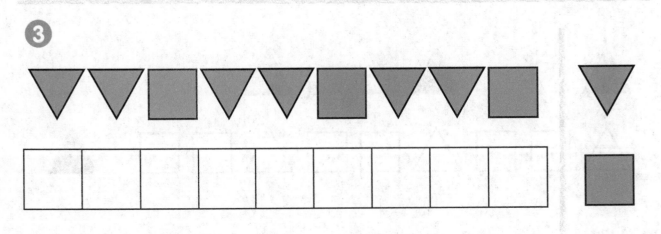

Directions: 1. Identify the pattern. Draw the shapes to copy the pattern. Trace the shape that could come next. Tell how you know. **2–3.** Identify the pattern. Draw the shapes to copy the pattern. Draw a circle around the shape that could come next.

Name ....

On My Own

4

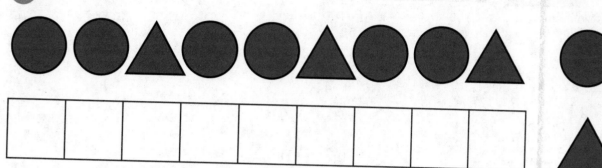

5

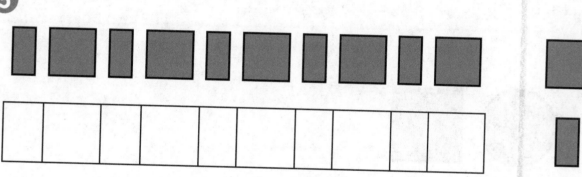

6

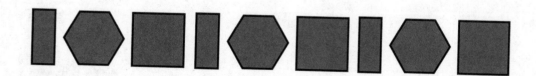

Directions: 4–5. Identify the pattern. Draw the shapes to copy the pattern. Draw a circle around the shape that could come next. Tell how you know. **6.** Identify the pattern. Draw the shape that could come next. Tell how you know.

Problem Solving

7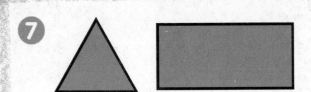

8

Directions: 7–8. Use the shapes given to create a pattern. Draw and color your pattern.

Time to color!

My Homework

Homework Helper

Need help? connectED.mcgraw-hill.com

1

2

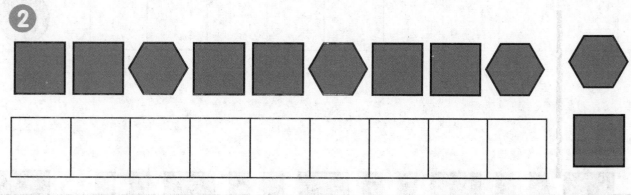

3

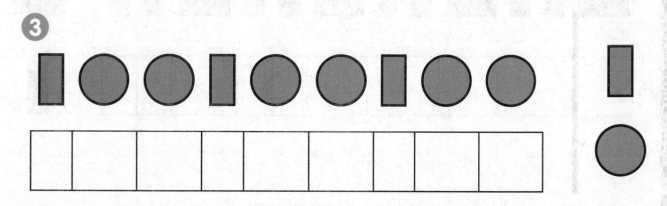

Directions: 1–3. Identify the pattern. Draw the shapes to copy the pattern. Circle the shape that could come next. Tell how you know.

4

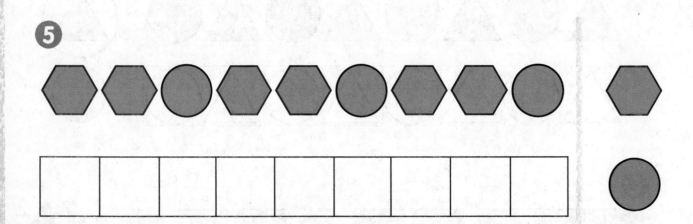

5

6

Directions: 4–6. Identify the pattern. Draw the shapes to copy the pattern.
Circle the shape that could come next. Tell how you know.

Math at Home Use shapes to make a pattern. Have your child copy the pattern
and tell what could come next in the pattern.

Geometry
K.G.1, K.G.2, K.G.5

CCSS

Shapes and Position

Lesson 6

ESSENTIAL QUESTION
How can I compare shapes?

 Explore and Explain

Teacher Directions: Use ▪ ▪ ▪ ○ ⬡ . Place an attribute block on each object that has a matching shape. Describe it. Circle the object that is *above* the desk. Draw a square picture *above* the bed. Draw a rectangle basket *in front of* the bed.

See and Show

1

2

Directions: 1. Circle the object that is *beside* the rectangle picture. Describe its shape. Draw an X on the object that is above the box. Describe its shape. **2.** Circle the object that is *above* the barn. Describe its shape. Draw an X on the object that is *in front* of the barn. Describe its shape.

Name

On My Own

What shape am I?

3

4

Directions: 3. Circle the object that is *below* the table. Describe its shape. Draw an X on the object that is *next to* the box. Describe its shape. Draw a box around the object that is *above* the box. Describe its shape. **4.** Circle the object that is *above* the clipboard. Describe its shape. Draw a box around the object that is *next to* the racket. Describe its shape.

Problem Solving

Directions: 5. Draw a window *above* the door and *next to* the other window. Describe its shape. **6.** Draw a swing *behind* the dog and *next to* the other swing. Describe its shape.

Name
_____

My Homework

Lesson 6

Shapes and Position

Homework Helper

eHelp

Need help? connectED.mcgraw-hill.com

1

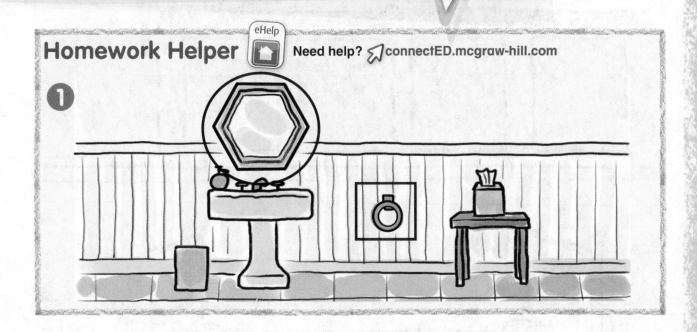

2

Directions: 1. Circle the object that is *above* the sink. Describe its shape. Draw a box around the object that is *next to* the tissue box. Describe its shape. **2.** Circle the object that is *beside* the keyboard. Describe its shape. Draw an X on the object that is *below* the table. Describe its shape.

Directions: 3. Circle the object that is *next to* the refrigerator. Describe its shape. Draw a box around the object that is *above* the sink. Describe its shape. Draw an X on the object that is *below* the sink. Describe its shape. **4.** Circle the object that is *next to* the basketball hoop. Describe its shape. Draw an X on the object that is *next to* the cone in the circle. Describe its shape.

Math at Home Find small circle, triangle, rectangle, hexagon, and square objects. Arrange them and have your child describe their position.

Geometry
K.G.2, K.G.5, K.G.6

CCSS

Compose New Shapes

Lesson 7

ESSENTIAL QUESTION
How can I compare shapes?

Explore and Explain

Tools | Watch

Teacher Directions: Use to make a large rectangle. Then make a large square. Trace one of the shapes you made.

See and Show

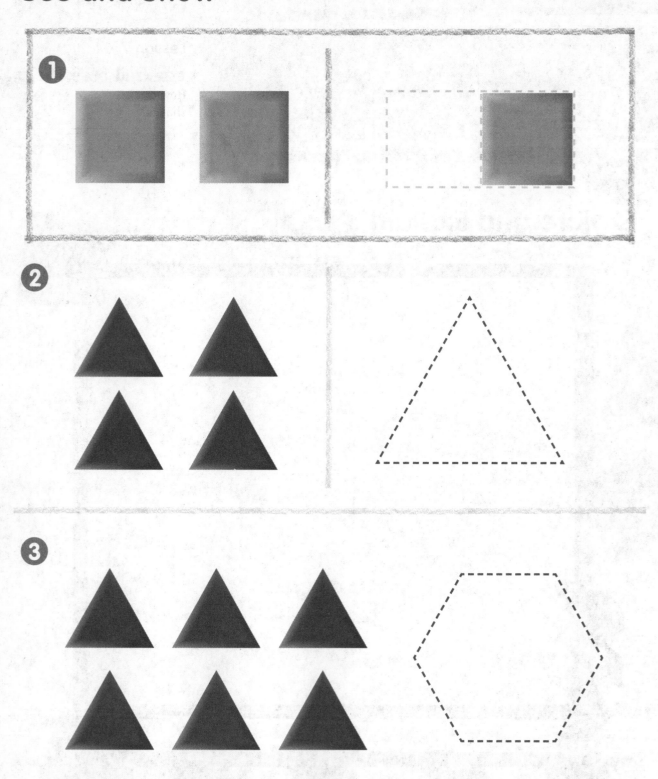

Directions: 1–3. Use the pattern blocks shown to make the larger shape. Trace the larger shape. Name the larger shape.

On My Own

4

5

6

 Directions: 4–5. Use pattern blocks shown to make the larger shape. Trace the larger shape. Name the larger shape. **6.** Use nine pattern blocks to make a larger square. Trace the shapes you used.

Problem Solving

7

8

Let's make shapes!

 Directions: 7. Look at the rectangle. Draw the shapes that make up the rectangle. **8.** Look at the triangles. Can two triangles make a rectangle? Try it using pattern blocks. Circle the triangles if yes. Put an X on the triangles if no.

Name

My Homework

Lesson 7

Compose New Shapes

Homework Helper eHelp Need help? connectED.mcgraw-hill.com

1

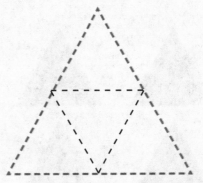

2

3

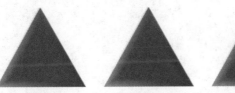
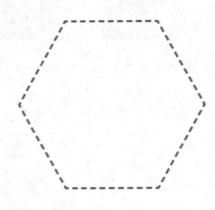

Directions: 1–3. Using *Manipulatives Masters: Pattern Blocks, Sheet 1*, color the pattern block triangles green and the squares orange. *Note: Parents, cut out the paper shapes for your child.* Use the cut out shapes to make the larger shape. Name the shape. Color it.

4

5

6

Copyright © The McGraw-Hill Companies, Inc.

 Directions: 4–6. Use the cut out shapes to make the the larger shape. Name the shape. Color it.

Math at Home Have your child count 8 cut out squares. Direct your child to make a rectangle with the squares.

Name _____

Problem Solving
STRATEGY: Use Logical Reasoning

What shapes are missing?

Use Logical Reasoning

Directions: Use pattern blocks to find the missing shapes. Name and describe the shapes in the larger shape. Trace the shapes.

 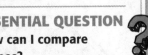

What shapes are missing?

Use Logical Reasoning

Directions: Use pattern blocks to find the missing shapes. Name and describe the shapes in the larger shape. Draw the shapes.

What shapes are missing?

Use Logical Reasoning

Directions: Use pattern blocks to find the missing shapes. Name and describe the shapes in the larger shape. Draw the shapes.

What shapes are missing?

Use Logical Reasoning

 Directions: Use pattern blocks to find the missing shapes. Name and describe the shapes in the larger shape. Draw the shapes.

My Homework

Lesson 8

Problem Solving: Use Logical Reasoning

What shapes are missing?

Use Logical Reasoning

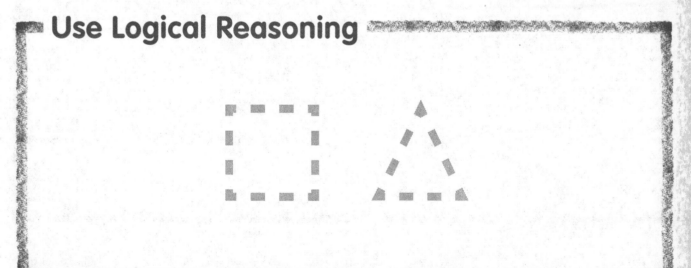

 Directions: Use patterns blocks from *Manipulative Masters: Pattern Blocks, Sheet I* to find the missing shapes. Name and describe the missing shapes. Trace the missing shapes. *Note: Parents, cut out the paper shapes for your child.*

What shapes are missing?

Use Logical Reasoning

Directions: Use pattern block cut outs to find the missing shapes. Name and describe the missing shapes. Draw the missing shapes.

Math at Home Take advantage of problem-solving opportunities during daily routines such as going to the park. Have your child identify shapes at the park.

Name

Model Shapes in the World

Explore and Explain Tools Watch

Teacher Directions: Use ■ ■ ● ▲ to make a new shape. Trace the shapes to show your new shape. Describe the new shape to a partner.

See and Show

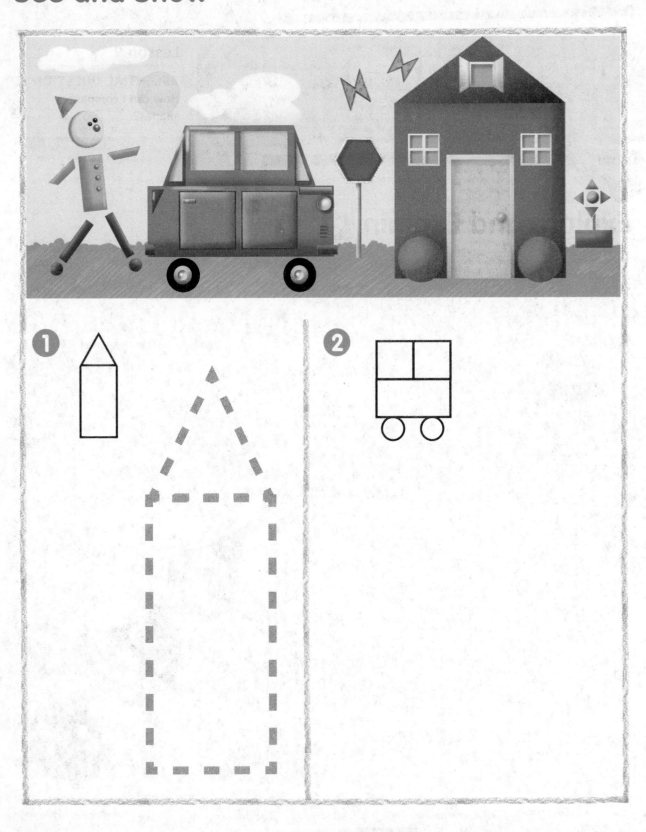

①

②

 Directions: 1–2. Look at the picture. What shapes make the picture? Name the shapes. Use attribute blocks to model the object in the picture. Trace the attribute blocks.

On My Own

3

4

 Directions: 3–4. Look at the picture. What shapes make the picture? Name the shapes. Use attribute blocks and pattern blocks to model the object in the picture. Trace the attribute blocks.

Online Content at **connectED.mcgraw-hill.com**   Chapter 11 • Lesson 9   675

Problem Solving

5

 Directions: 5. Use square, rectangle, circle, triangle, and hexagon pattern blocks and attribute blocks to make fish in the bowl. Trace the shapes and color them. Tell a partner the names of the shapes you used.

Name ....

My Homework

Lesson 9

Model Shapes in the World

Homework Helper

eHelp

Need help? connectED.mcgraw-hill.com

①

②

Directions: 1–2. Look at the object. What shapes make the object? Name the shapes. Draw the shapes used to model the object in the picture. Color the shapes.

3

4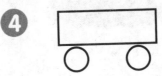

Directions: 3–4. Look at the picture. What shapes make the picture? Name the shapes. Draw the shapes used to model the object in the picture. Color the shapes.

Math at Home Choose another object in the picture above. Have your child draw the shapes in the picture above to model the object.

Name

My Review

Vocabulary Check

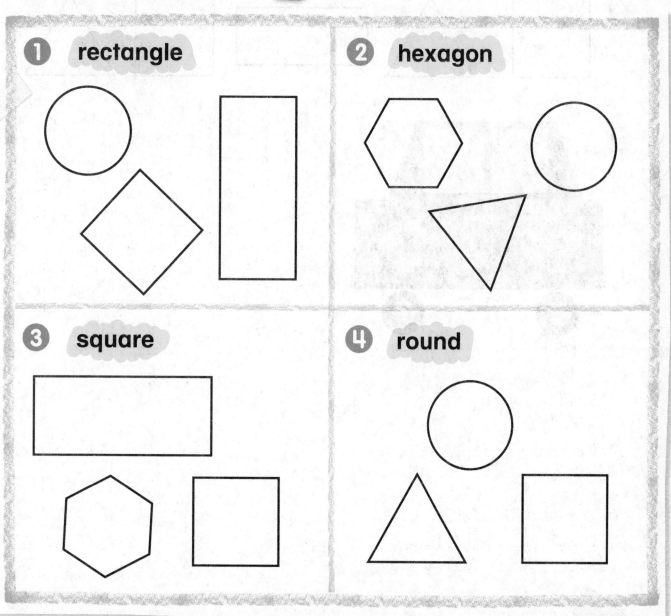

1 rectangle

2 hexagon

3 square

4 round

 Directions: 1. Color the rectangle red. **2.** Color the hexagon blue.
3. Color the square orange. **4.** Color the shape that is round purple.

Concept Check

5

6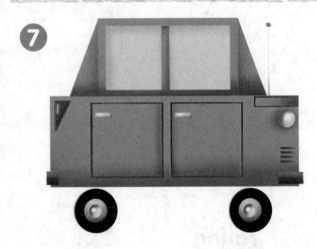

7

Directions: 5–6. Name the shape. Describe it. Compare the shape to each shape in the group. Circle the matching shapes. **7.** Look at the picture. What shapes do you see? Use attribute blocks to model the car. Trace the shapes.

Problem Solving

HAPPY DAY!

Directions: 8. Look at the picture. Circle the object that is *above* the tree. Describe its shape. Draw an X on the object that is *next to* the tree. Describe its shape. Draw a wagon *in front* of the girls. Describe its shape.

sides   vertices

sides   vertices

sides   vertices

sides   vertices

 Directions: Trace each shape. Identify and describe the shapes. Write how many sides and vertices.

ESSENTIAL QUESTION

How do I identify and compare three-dimensional shapes?

My Dreams Take Me Places!

Watch a video!

Watch

My Common Core State Standards

Geometry

K.G.1 Describe objects in the environment using the names of shapes, and describe the relative position of these objects using terms such as above, below, in front of, behind, and next to.

K.G.2 Correctly name shapes regardless of their orientations or overall size.

K.G.3 Identify shapes as two-dimensional (lying in a plane, "flat") or three-dimensional ("solid").

K.G.4 Analyze and compare two- and three-dimensional shapes, in different sizes and orientations, using informal language to describe their similarities, differences, parts (e.g., number of sides and vertices/"corners") and other attributes (e.g., having sides of equal length).

K.G.5 Model shapes in the world by building shapes from components (e.g., sticks and clay balls) and drawing shapes.

Standards for Mathematical PRACTICE

1. Make sense of problems and persevere in solving them.
2. Reason abstractly and quantitatively.
3. Construct viable arguments and critique the reasoning of others.
4. Model with mathematics.
5. Use appropriate tools strategically.
6. Attend to precision.
7. Look for and make use of structure.
8. Look for and express regularity in repeated reasoning.

= focused on in this chapter

Name

 Check ✓ ← Go online to take the Readiness Quiz

1

2

3

4

 Directions: 1–4. Circle the shapes in the row that are the same.

My Math Words

Review Vocabulary

 Directions: Trace each word and tell its meaning. Look at the shapes that make the objects in the clouds. Color the squares red. Draw an X on each circle. Tell how you know.

My Vocabulary Cards

 Vocab abc

Mathematical **PRACTICE**

cone
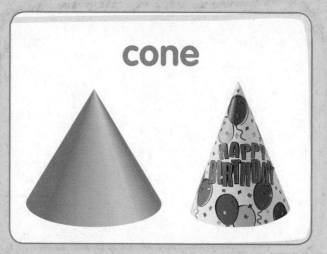

cube

cylinder

roll

slide

sphere

cube

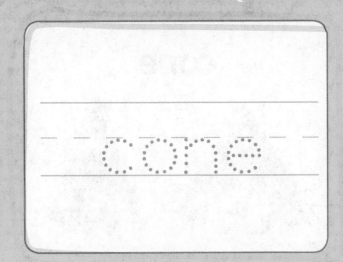

cone

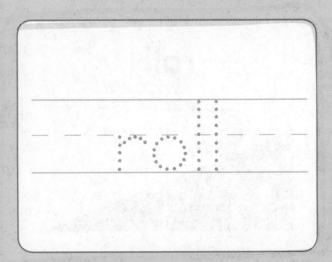

roll

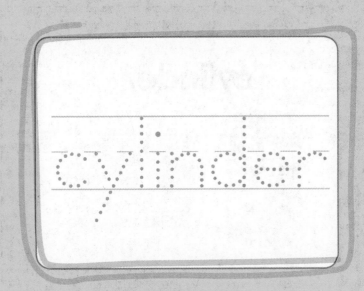

cylinder

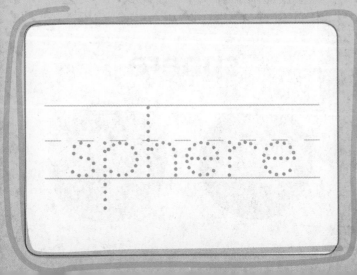

sphere

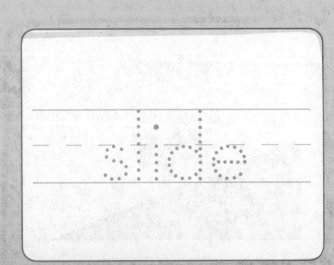

slide

stack

Teacher Directions:
More ideas for Use

• Direct a student to act out *roll*, *stack*, or *slide* with a partner. Have another student find the vocabulary word that indicates the action acted out.

• Instruct students to use the blank cards to create alphabet cards. Suggest that students write a letter on the front of a card and then write a math word that begins with that letter on the back.

stock

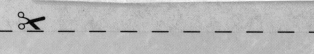

My Foldable

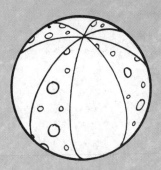

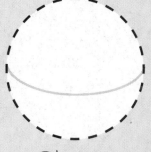 slide roll stack

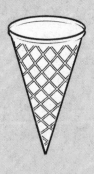

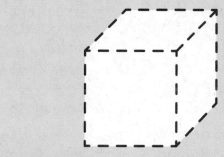 slide roll stack

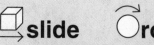 slide  roll stack

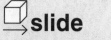 slide   roll stack

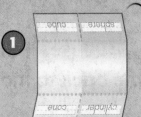
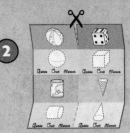
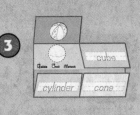
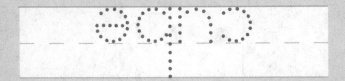

cube

sphere

cone

cylinder

Name _____

Spheres and Cubes →

Lesson 1

ESSENTIAL QUESTION
How do I identify
and compare three-
dimensional shapes?

Explore and Explain

 Teacher Directions: Go on a three-dimensional shape walk around the classroom. Identify objects that are shaped like spheres and cubes. Draw a picture of one of the objects beside the matching shape. Tell whether the objects are solid or flat.

See and Show

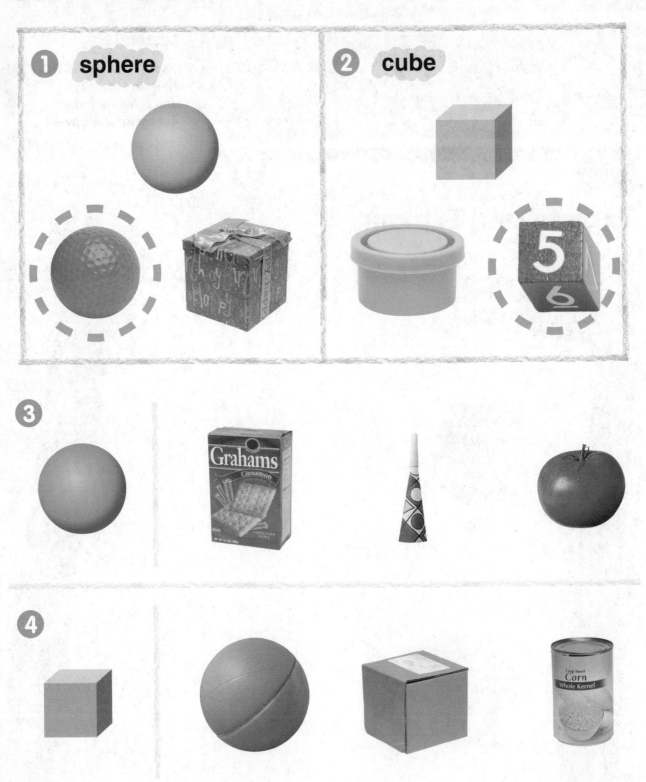

① sphere

② cube

③

④

Directions: 1–2. Name the shape above the objects. Describe it. Compare it to the shapes of the objects below it. Trace the circle around the matching shape. **3–4.** Name the first shape in the row. Describe it. Compare it to the shapes of the objects in the row. Circle the matching shape.

Name _____

On My Own

5

6

7

8

 Directions: 5–8. Name the first shape in the row. Describe it. Compare it to the shapes of the objects in the row. Circle the matching shape.

 Directions: 9. Name the shape of the planet. Find those shapes on the page. Draw lines from those shapes to the planet. Identify the part of the rocket that is shaped like a cube. Find those shapes on the page. Draw lines from those shapes to the rocket.

My Homework

Homework Helper

 eHelp

Need help? connectED.mcgraw-hill.com

1

2

3

4

 Directions: 1–4. Name the first shape in the row. Describe it. Compare it to the shapes of the objects in the row. Circle the matching shape.

⑤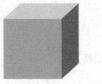

⑥

Vocabulary Check

⑦ sphere

⑧ cube

 Directions: 5–6. Name the first shape in the row. Describe it. Compare it to the shapes of the objects in the row. Circle the matching shape. **7.** Draw an X on the objects that are not shaped like a sphere. **8.** Draw an X on the objects that are not shaped like a cube.

Math at Home Discuss the spheres and cubes shown on the My Homework pages. Help your child create a list of objects shaped like spheres and cubes and draw the objects.

Cylinders and Cones

Lesson 2

ESSENTIAL QUESTION
How do I identify and compare three-dimensional shapes?

 Explore and Explain

I like cones!

 Teacher Directions: Go on a three-dimensional shape walk around the classroom. Identify objects that are shaped like cylinders and cones. Draw a picture of one of the objects beside the matching shape. Tell whether the objects are solid or flat.

See and Show

1 cylinder

2 cone

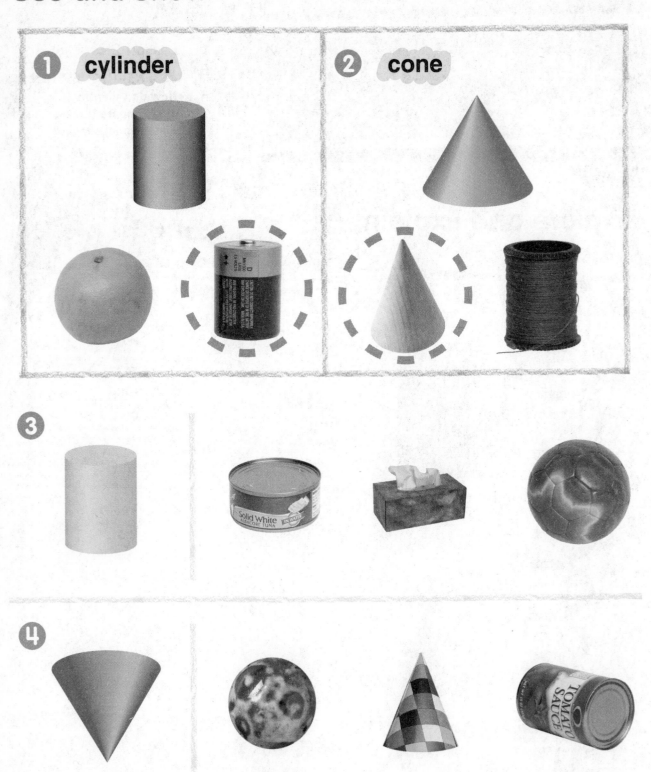

3

4

Directions: 1–2. Name the shape above the objects. Describe it. Compare it to the shapes of the objects below it. Trace the circle around the matching shape. **3–4.** Name the first shape in the row. Describe it. Compare it to the shapes of the objects in the row. Circle the matching shape.

On My Own

5

6

7

8

 Directions: 5–8. Name the first shape in the row. Describe it. Compare it to the shapes of the objects in the row. Circle the matching shape.

 Directions: 9. Point to the broken pieces of crayons. Name the shape of each piece. Identify each shape by coloring the cones blue and the cylinders red.

Name ....

My Homework

Lesson 2

Cylinders and Cones

Homework Helper eHelp Need help? connectED.mcgraw-hill.com

1

2

3

4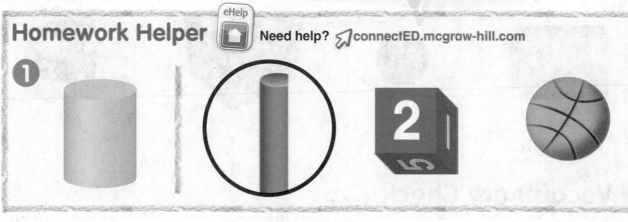

 Directions: 1–4. Name the first shape in the row. Describe it. Compare it to the shapes of the objects in the row. Circle the matching shape.

⑤

⑥

Vocabulary Check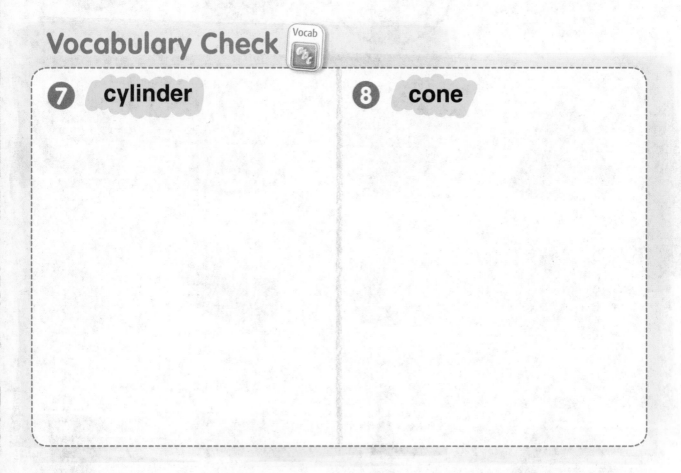

⑦ cylinder

⑧ cone

 Directions: 5–6. Name the first shape in the row. Describe it. Compare it to the shapes of the objects in the row. Circle the matching shape. **7.** Draw an object that is shaped like a cylinder. Tell what you drew. **8.** Draw an object that is shaped like a cone. Tell what you drew.

Math at Home Have your child find objects in your home shaped like cylinders and cones.

Compare Solid Shapes

Lesson 3

ESSENTIAL QUESTION
How do I identify and compare three-dimensional shapes?

Explore and Explain

Let's roll!

Teacher Directions: Use ●, ■, █, and ▲ to identify and compare solid shapes. Show which shapes roll, slide, and can have another shape stacked on it. Name an object in the classroom that rolls, slides, or stacks the same as one of the solid shapes. Draw the object. Tell how you know.

See and Show

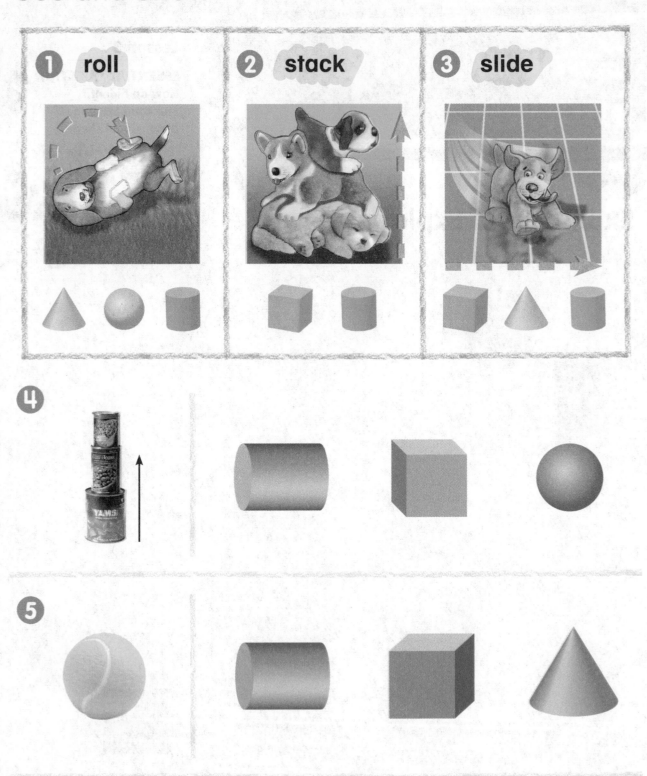

1 roll

2 stack

3 slide

4

5

Directions: 1–3. Trace the arrows to show roll, stack, and slide. Name the shapes shown that roll, stack, and slide. **4.** Identify the shape of the first objects in the row. Describe it. Circle the other shape(s) in the row that stack in the same way. **5.** Identify the shape of the first object in the row. Describe it. Circle the other shape(s) in the row that roll.

Name

I slide, too!

On My Own

6

7

8

9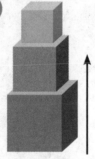

 Directions: 6. Identify the shape of the first object in the row. Describe it. Circle the other shape(s) in the row that slide. **7.** Identify the shape of the first object in the row. Describe it. Circle the other shape(s) in the row that roll. **8.** Draw an object that would roll as a cylinder. **9.** Draw an object that would stack as the cubes are stacked.

 Directions: 10. Look at the shapes on the carpet. Find a shape that rolls, stacks, and slides. Color it orange. Find a shape that only slides and rolls and cannot have a shape stacked on it. Color it purple. Find a shape that only slides and stacks. Color it yellow. Find a shape that only rolls. Color it blue.

My Homework

Homework Helper

eHelp

Need help? connectED.mcgraw-hill.com

1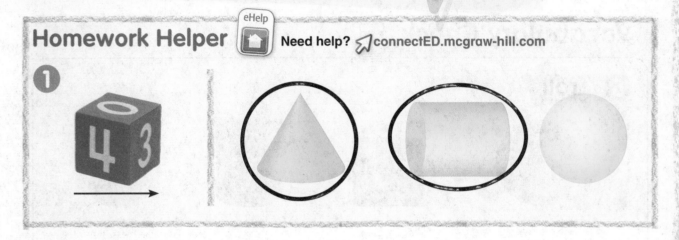

2

3

 Directions: 1. Identify the shape of the first object in the row. Describe it. Circle the other shape(s) in the row that slide. **2.** Identify the shape of the first objects in the row. Describe it. Circle the other shape(s) in the row that stack in the same way. **3.** Identify the shape of the first object in the row. Describe it. Circle the other shape(s) in the row that roll.

4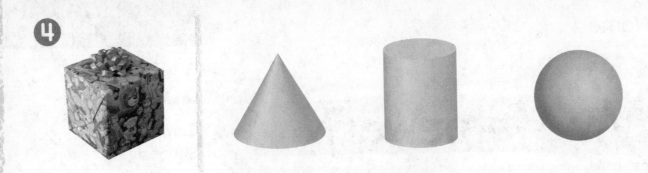

Vocabulary Check

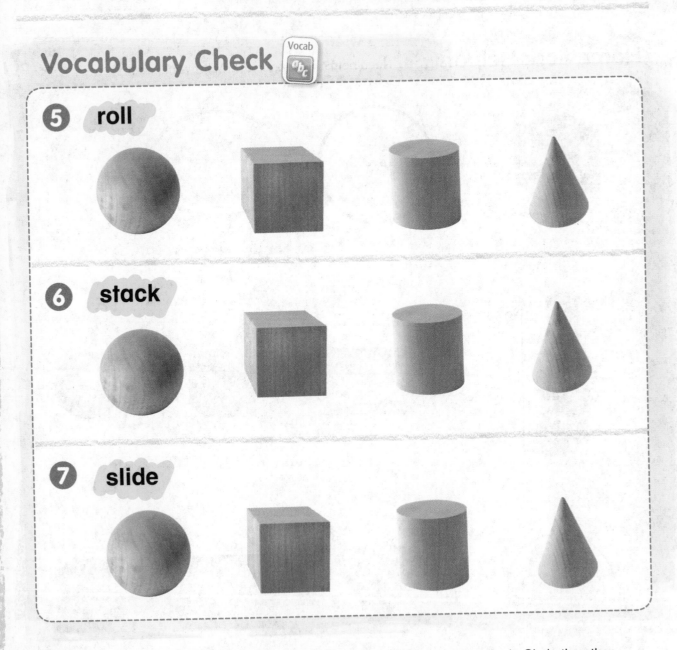

5 roll

6 stack

7 slide

Copyright © The McGraw-Hill Companies, Inc. (l to r, t to b) Royalty-Free/CORBIS; (2, 3, 4) The McGraw-Hill Companies

Directions: 4. Identify the shape of the first shape in the row. Describe it. Circle the other shape(s) in the row that slide. **5.** Identify the shapes. Circle the shapes that roll. **6.** Identify the shapes. Circle the shapes that can have another shape stacked on it. **7.** Identify the shapes. Circle the shapes that slide.

Math at Home Have your child name the solid shapes on the My Homework pages that roll, stack, and slide. Have your child imitate these movements.

Name _____

Vocabulary Check

1 sphere

2 cylinder

Concept Check

3

 Directions: 1. Draw an X on the objects that are not shaped like a sphere. **2.** Draw an X on the objects that are not shaped like a cylinder. **3.** Name the first shape in the row. Compare it to the shapes of the objects in the row. Circle the matching shape.

4

5

6

7

Directions: 4. Name the first shape in the row. Compare it to the shapes of the objects in the row. Circle the matching shape. **5.** Identify the first objects in the row. Describe them. Circle the other shape(s) in the row that stack in the same way. **6.** Identify the first object in the row. Describe it. Circle the other shape(s) in the row that roll. **7.** Identify the first object in the row. Describe it. Circle the other shape(s) in the row that slide.

Name _____

Problem Solving
STRATEGY: Act It Out

Lesson 4

ESSENTIAL QUESTION
How do I identify and compare three-dimensional shapes?

What will stack on the cube?

Act It Out

 Teacher Directions: Name and describe the shapes in the castle. Trace the circles to show the shapes that you could stack on the cube. Use shapes to check the answer.

What will stack on the cylinder?

Act It Out

Directions: Name and describe the shapes in the tower. Circle the shape(s) that you could stack on the cylinder. Draw an X on the shape(s) that could not stack on the cylinder. Use shapes to check the answer.

What will stack on the cone?

Act It Out

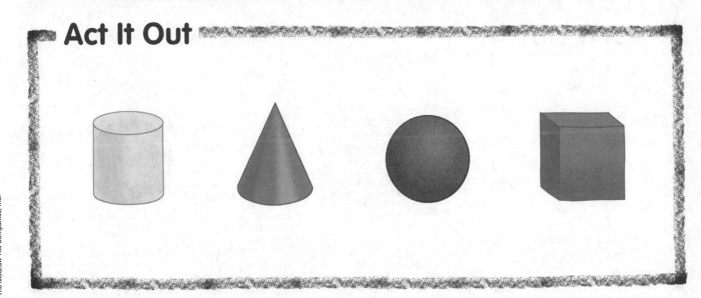

 Directions: Name and describe the shapes in the tower. Circle the shape(s) that you could stack on the cone. Draw an X on the shape(s) that could not stack on the cone. Use shapes to check your answer.

What will stack on the cube?

Act It Out

Directions: Name and describe the shapes in the tower. Circle the shape(s) that you could use to stack on the cube. Draw an X on the shape(s) that could not stack on the cube. Use shapes to check your answer.

Geometry
K.G.2, K.G.4

CCSS

My Homework

What will stack on the cylinder?

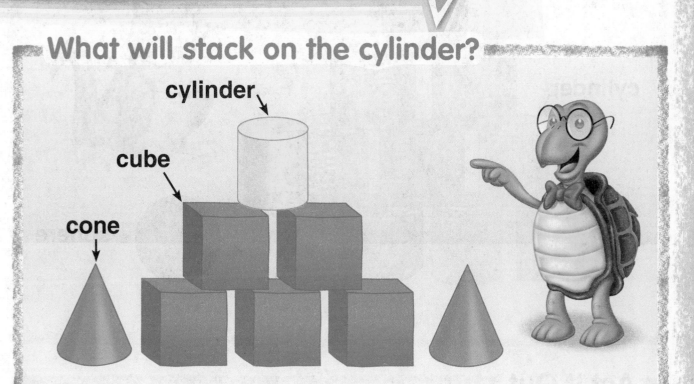

cylinder

cube

cone

Act It Out

Directions: Name and describe the shapes in the tower. Trace the circles to show the shapes that you could stack on the cylinder. Use a canned good, cube shaped tissue box, and ball to check the answer.

What will stack on the cube?

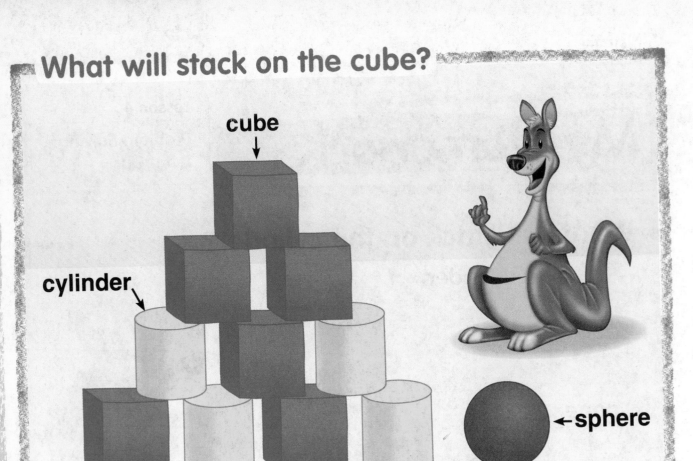

cube

cylinder

←sphere

Act It Out

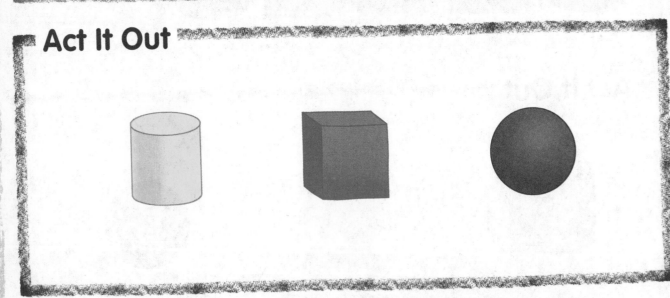

Directions: Name and describe the shapes in the tower. Circle the shapes that you could stack on the cube. Use a canned good, cube shaped tissue box, and ball to check the answer.

Math at Home Take advantage of problem-solving opportunities during daily routines such as playing with blocks, cleaning up toys, or putting away groceries to identify objects that can stack, roll, or slide.

Geometry
K.G.1, K.G.2, K.G.3, K.G.4, K.G.5

CCSS

Model Solid Shapes in Our World

Lesson 5

ESSENTIAL QUESTION
How do I identify and compare three-dimensional shapes?

Explore and Explain Watch ▶

 Teacher Directions: Circle all the cones, cylinders, cubes, and spheres in the picture. Tell a classmate the name of each shape you found. Describe the shapes. Find objects in the classroom that are the same shapes.

See and Show

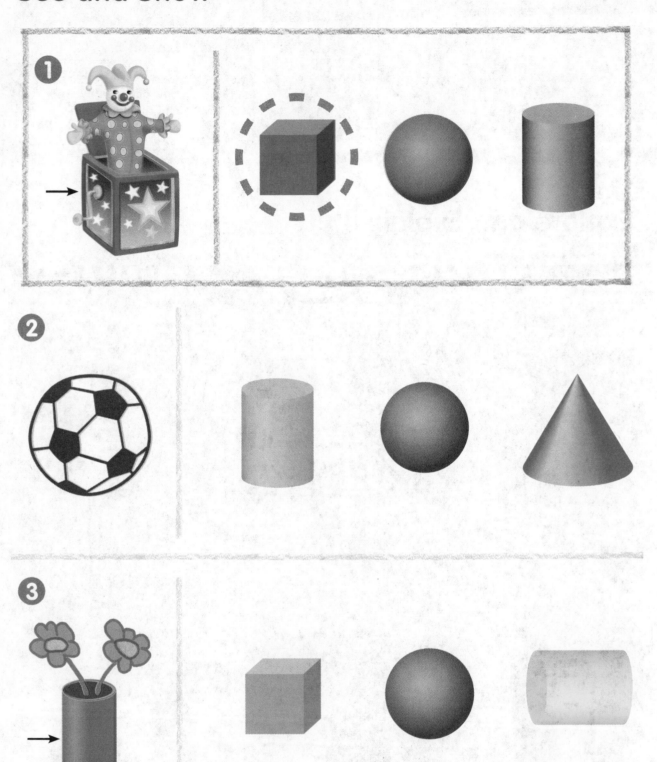

1

2

3

Directions: 1–3. Name the shape of the first object in the row. Describe it. Compare it
to each shape in the row. Circle the matching shape.

720   Chapter 12 • Lesson 5

Name _____

On My Own

④

⑤

⑥

 Directions: 4–6. Name the shape of the first object in the row. Describe it. Compare it to each shape in the row. Circle the matching shape.

Online Content at ⇗ **connectED.mcgraw-hill.com**

Chapter 12 • Lesson 5   721

Problem Solving

 Directions: 7. Color all the cubes yellow. Color all the cones red. Color all the spheres blue. Color all the cylinders orange. Describe the shapes.

Name ....

My Homework

Lesson 5

Model Solid Shapes in Our World

Homework Helper

eHelp

Need help? connectED.mcgraw-hill.com

1

2

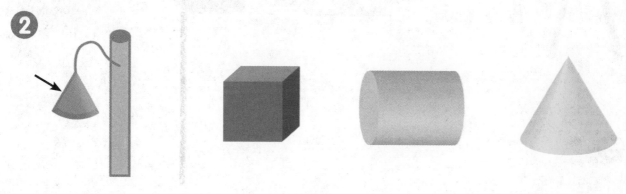

3

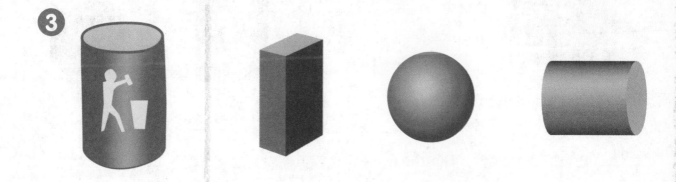

Directions: 1–3. Name the shape of the first object in the row. Describe it. Compare it to each shape in the row. Circle the matching shape.

4

5

6

Directions: 4–6. Name the shape of the first object in the row. Describe it. Compare it to each shape in the row. Circle the matching shape.

Math at Home Take a walk outside. Look for three-dimensional shapes. Draw pictures of the objects that are the shapes of cones, spheres, cylinders, and cubes.

Name

Vocabulary Check

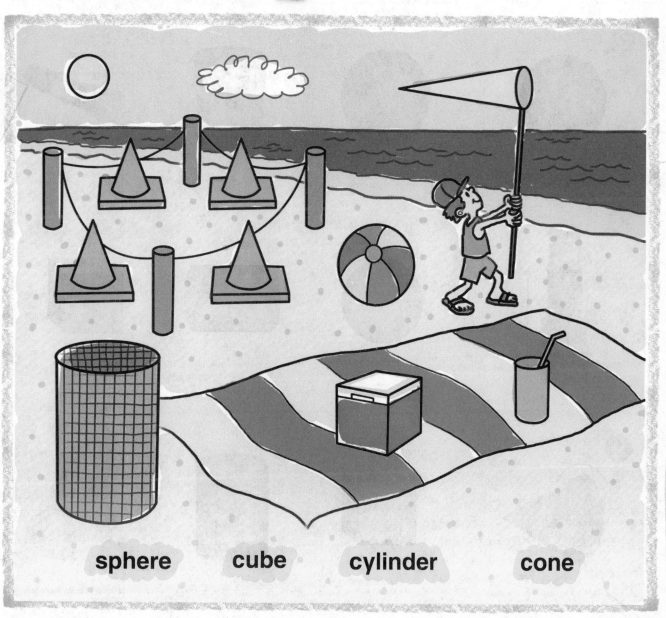

sphere   **cube**   **cylinder**   **cone**

 Directions: Draw a line from each vocabulary word to a shape that matches the word. Draw a box around the objects shaped like a sphere. Draw an X on the object shaped like a cube. Draw a line under the objects shaped like a cylinder. Circle the objects shaped like a cone.

Concept Check

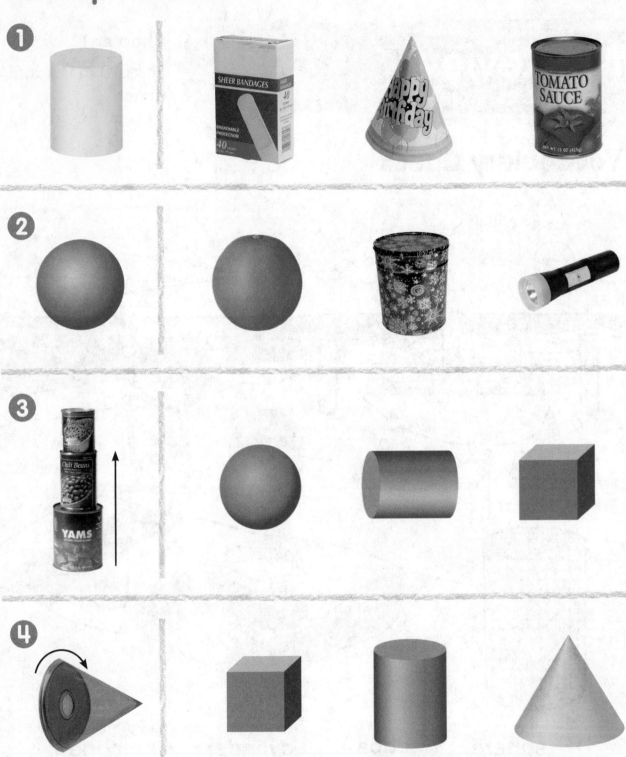

 Directions: 1–2. Name the first shape in the row. Compare it to the shapes of the objects in the row. Circle the matching shape. **3.** Name the shape of the first objects in the row. Circle the other shapes in the row that stack in the same way. **4.** Name the shape of the first object in the row. Circle the other shapes in the row that roll.

Name

 Problem Solving

 Directions: 5. Name, identify, and describe the shapes. Color the spheres yellow. Color the cubes orange. Color the cones blue. Color the cylinders red.

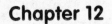

Chapter 12

ESSENTIAL QUESTION
How do I identify
and compare three-
dimensional shapes?

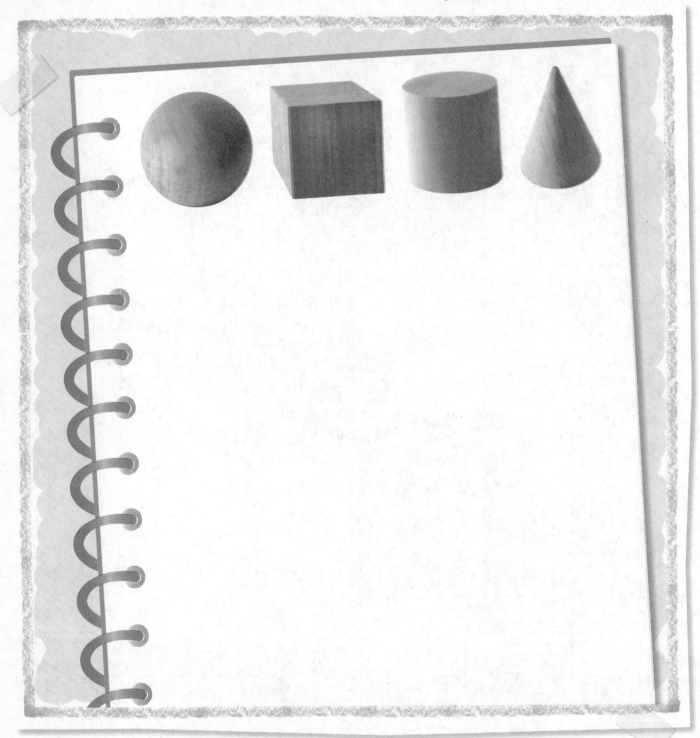

Directions: Name each solid shape. Describe the shapes. Circle one of the shapes. Tell if the shape is solid or flat. Draw an object that is shaped the same as that shape. Describe the object to a classmate. Ask the classmate to name the shape of the object.

Glossary/Glosario

 Vocab → Go online for the eGlossary.

Aa	**English**	**Spanish/Español**

above

above

encima

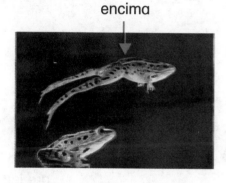

encima

add

 **= 5**

3 ducks   2 more join   5 ducks in all

sumar

3 patos   se unen 2 más   5 patos en total

afternoon

UNION SCHOOL DISTRICT 3

tarde

UNION SCHOOL DISTRICT 3

Aa

alike (same)

alike   different

igual

iguales   diferentes

are left

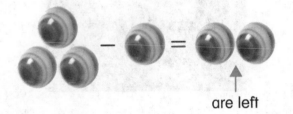

are left

quedan

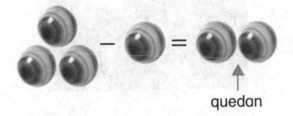

quedan

Bb

behind

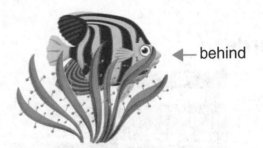

← behind

detrás

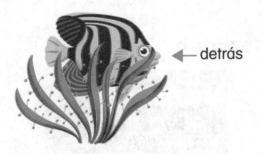

← detrás

below

below

debajo

debajo

beside

The cat is beside the dog.

al lado

El gato está al lado del perro.

calendar

			April			
Sunday	Monday	Tuesday	Wednesday	Thursday	Friday	Saturday
		1	2	3	4	5
6	7	8	9	10	11	12
13	14	15	16	17	18	19
20	21	22	23	24	25	26
27	28	29	30			

calendario

			abril			
domingo	lunes	martes	miércoles	jueves	viernes	sábado
		1	2	3	4	5
6	7	8	9	10	11	12
13	14	15	16	17	18	19
20	21	22	23	24	25	26
27	28	29	30			

Cc

capacity

holds more   holds less

capacidad

contiene más   contiene menos

circle

círculo

compare

← more than

← less than

comparar

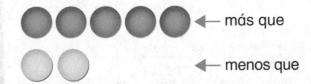

← más que

← menos que

cone

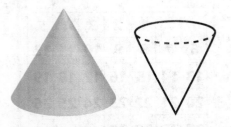

cono

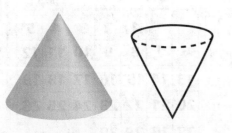

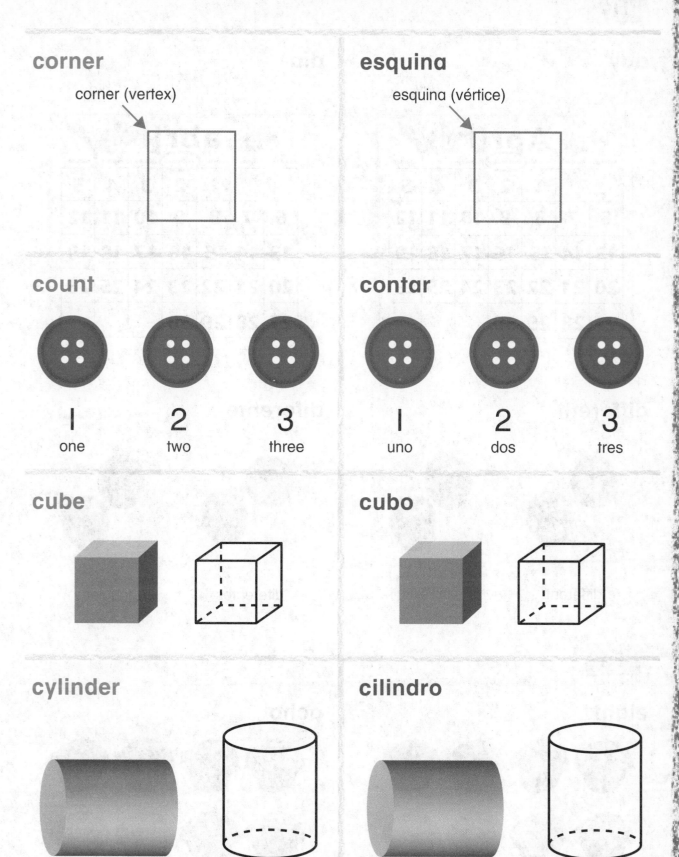

corner

corner (vertex)

esquina

esquina (vértice)

count

1   one
2   two
3   three

contar

1   uno
2   dos
3   tres

cube

cubo

cylinder

cilindro

day

day

día

día

different

different   alike

diferente

diferentes   iguales

eight

ocho

eighteen

dieciocho

eleven

once

equals sign (=)

$$4 + 1 = 5$$

↑
equals

signo igual (=)

$$4 + 1 = 5$$

↑
es igual a

equal to

igual a

Ee

evening

noche

fifteen

quince

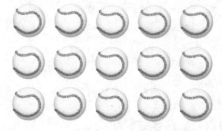

five

cinco

four

cuatro

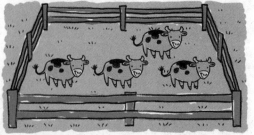

fourteen

catorce

Gg

greater than

mayor que

Hh

heavy (heavier)

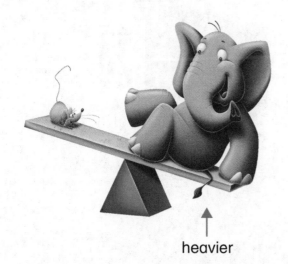

heavier

pesado (más pesado)

más pesado

Hh

height

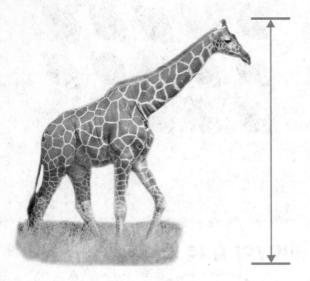

altura

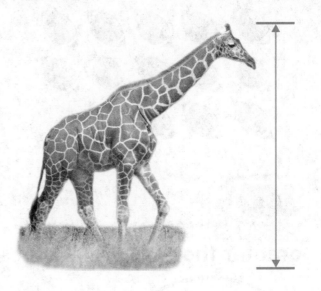

hexagon

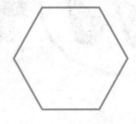

hexágono

holds less

holds less

contiene menos

contiene menos

holds more

↑

holds more

contiene más

↑

contiene más

holds the same

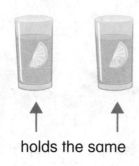

↑   ↑

holds the same

contiene la misma cantidad

↑   ↑

contiene la misma cantidad

in all

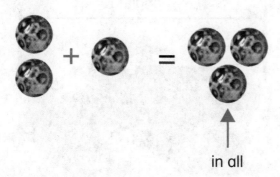

↑

in all

en total

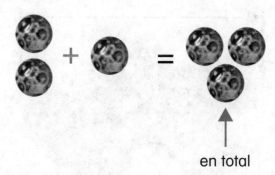

↑

en total

li

in front of

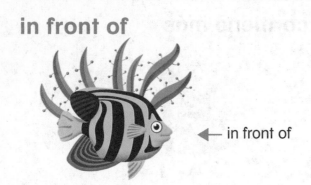

← in front of

en frente de

← en frente de

Jj

join

3 birds and 2 birds join.

juntar

Hay 3 aves y se les juntan 2 más.

Ll

length

length

longitud

longitud

less than

menor que

light (lighter)

lighter

liviano (más liviano)

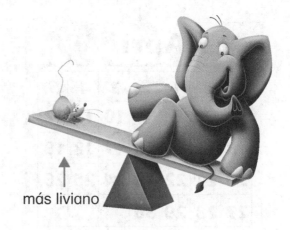

más liviano

long (longer)

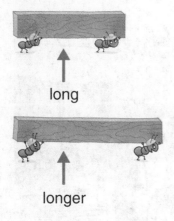

long

longer

largo (más largo)

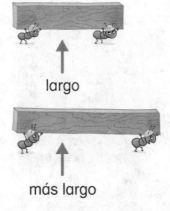

largo

más largo

minus sign (−)

$$5 - 2 = 3$$

↑ minus

signo menos (−)

$$5 - 2 = 3$$

↑ menos

month

month →

mes

mes →

morning

mañana

next to

The cat is next to the dog.

junto a (al lado)

El gato está junto al perro.

nine

nueve

nineteen

diecinueve

number

3

tells how many

número

3

dice cuántos hay

one

uno

ordinal numbers

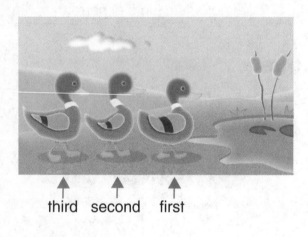

third   second   first

números ordinales

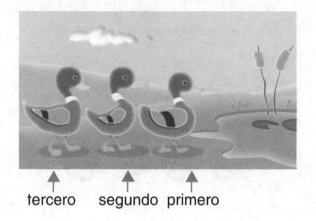

tercero   segundo   primero

pattern

A, B, A, B, A, B

repeating pattern

patrón

A, B, A, B, A, B

patrón que se repite

plus sign (+)

$$5 + 2 = 7$$

↑

plus

signo más (+)

$$5 + 2 = 7$$

↑

más

position

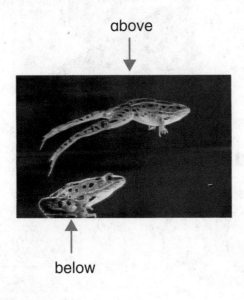

above

below

posición

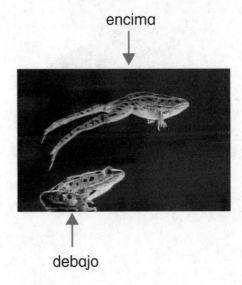

encima

debajo

rectangle

rectángulo

Rr

repeating pattern

repeating pattern

patrón que se repite

patrón que se repite

roll

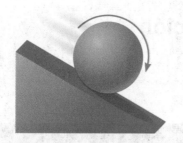

rodar

round

round   not round

redondo

redondo   no redondo

Ss

same number

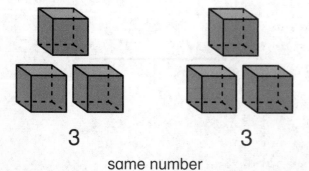

3   3

same number

el mismo número

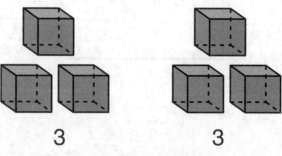

3   3

el mismo número

separate

separar

seven

siete

seventeen

diecisiete

shape

figura

Ss

short (shorter)	corto (más corto)

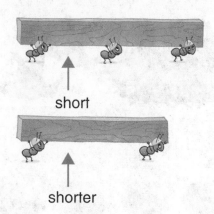

short

shorter

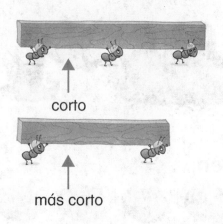

corto

más corto

side	lado

side →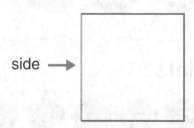

lado →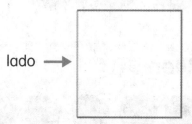

six	seis

sixteen	dieciséis

size

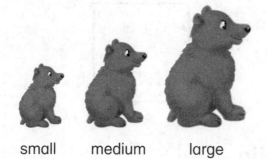

small   medium   large

tamaño

pequeño   mediano   grande

slide

deslizar

sort

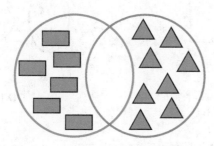

sorted or grouped by shape

ordenar

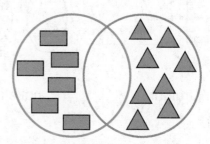

ordenados o agrupados por su forma

sphere

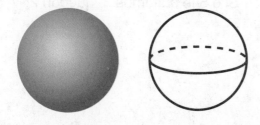

esfera

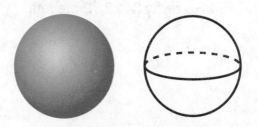

Ss

square 	**cuadrado** 
stack 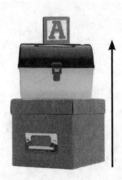	**pila** 
straight  straight not straight	**recto** 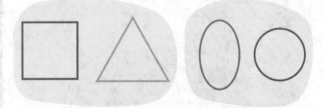 recto no recto
subtract (subtraction)  5 take away 3 is 2. 2 are left.	**restar (resta)**  Si a 5 le quitamos 3, quedan 2.

tall (taller)

taller

alto (más alto)

más alto

ten

diez

thirteen

trece

three

tres

Tt

three-dimensional shape

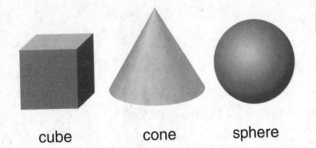

cube   cone   sphere

figura tridimensional

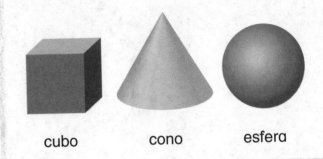

cubo   cono   esfera

today

yesterday   today

Sunday	Monday	Tuesday	Wednesday	Thursday	Friday	Saturday
		1	2	3	4	5
6	7	8	9	10	11	12
13	14	15	16	17	18	19
20	21	22	23	24	25	26
27	28	29	30			

hoy

ayer   hoy

domingo	lunes	martes	miércoles	jueves	viernes	sábado
		1	2	3	4	5
6	7	8	9	10	11	12
13	14	15	16	17	18	19
20	21	22	23	24	25	26
27	28	29	30			

tomorrow

today   tomorrow

Sunday	Monday	Tuesday	Wednesday	Thursday	Friday	Saturday
		1	2	3	4	5
6	7	8	9	10	11	12
13	14	15	16	17	18	19
20	21	22	23	24	25	26
27	28	29	30			

mañana

hoy   mañana

domingo	lunes	martes	miércoles	jueves	viernes	sábado
		1	2	3	4	5
6	7	8	9	10	11	12
13	14	15	16	17	18	19
20	21	22	23	24	25	26
27	28	29	30			

triangle

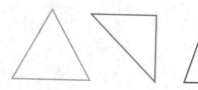

triángulo

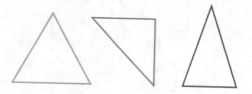

twelve

doce

twenty

veinte

two

dos

two-dimensional shape

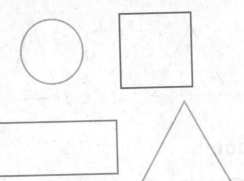

figura bidimensional

vertex

vertex
(corner)

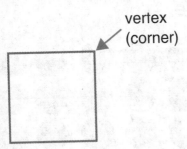

vértice

vértice
(esquina)

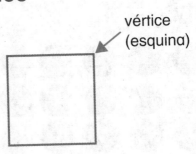

week

week

semana

semana

weight

heavy   light

peso

pesado   liviano

Yy

year

año

January

S	M	T	W	T	F	S
						1
2	3	4	5	6	7	8
9	10	11	12	13	14	15
16	17	18	19	20	21	22
23	24	25	26	27	28	29
30	31					

February

S	M	T	W	T	F	S
		1	2	3	4	5
6	7	8	9	10	11	12
13	14	15	16	17	18	19
20	21	22	23	24	25	26
27	28					

March

S	M	T	W	T	F	S
		1	2	3	4	5
6	7	8	9	10	11	12
13	14	15	16	17	18	19
20	21	22	23	24	25	26
27	28	29	30	31		

April

S	M	T	W	T	F	S
					1	2
3	4	5	6	7	8	9
10	11	12	13	14	15	16
17	18	19	20	21	22	23
24	25	26	27	28	29	30

May

S	M	T	W	T	F	S
1	2	3	4	5	6	7
8	9	10	11	12	13	14
15	16	17	18	19	20	21
22	23	24	25	26	27	28
29	30	31				

June

S	M	T	W	T	F	S
			1	2	3	4
5	6	7	8	9	10	11
12	13	14	15	16	17	18
19	20	21	22	23	24	25
26	27	28	29	30		

July

S	M	T	W	T	F	S
					1	2
3	4	5	6	7	8	9
10	11	12	13	14	15	16
17	18	19	20	21	22	23
24	25	26	27	28	29	30
31						

August

S	M	T	W	T	F	S
	1	2	3	4	5	6
7	8	9	10	11	12	13
14	15	16	17	18	19	20
21	22	23	24	25	26	27
28	29	30	31			

September

S	M	T	W	T	F	S
				1	2	3
4	5	6	7	8	9	10
11	12	13	14	15	16	17
18	19	20	21	22	23	24
25	26	27	28	29	30	

October

S	M	T	W	T	F	S
						1
2	3	4	5	6	7	8
9	10	11	12	13	14	15
16	17	18	19	20	21	22
23	24	25	26	27	28	29
30	31					

November

S	M	T	W	T	F	S
		1	2	3	4	5
6	7	8	9	10	11	12
13	14	15	16	17	18	19
20	21	22	23	24	25	26
27	28	29	30			

December

S	M	T	W	T	F	S
				1	2	3
4	5	6	7	8	9	10
11	12	13	14	15	16	17
18	19	20	21	22	23	24
25	26	27	28	29	30	31

enero

d	l	m	m	j	v	s
						1
2	3	4	5	6	7	8
9	10	11	12	13	14	15
16	17	18	19	20	21	22
23	24	25	26	27	28	29
30	31					

febrero

d	l	m	m	j	v	s
		1	2	3	4	5
6	7	8	9	10	11	12
13	14	15	16	17	18	19
20	21	22	23	24	25	26
27	28					

marzo

d	l	m	m	j	v	s
		1	2	3	4	5
6	7	8	9	10	11	12
13	14	15	16	17	18	19
20	21	22	23	24	25	26
27	28	29	30	31		

abril

d	l	m	m	j	v	s
					1	2
3	4	5	6	7	8	9
10	11	12	13	14	15	16
17	18	19	20	21	22	23
24	25	26	27	28	29	30

mayo

d	l	m	m	j	v	s
1	2	3	4	5	6	7
8	9	10	11	12	13	14
15	16	17	18	19	20	21
22	23	24	25	26	27	28
29	30	31				

junio

d	l	m	m	j	v	s
			1	2	3	4
5	6	7	8	9	10	11
12	13	14	15	16	17	18
19	20	21	22	23	24	25
26	27	28	29	30		

julio

d	l	m	m	j	v	s
					1	2
3	4	5	6	7	8	9
10	11	12	13	14	15	16
17	18	19	20	21	22	23
24	25	26	27	28	29	30
31						

agosto

d	l	m	m	j	v	s
	1	2	3	4	5	6
7	8	9	10	11	12	13
14	15	16	17	18	19	20
21	22	23	24	25	26	27
28	29	30	31			

septiembre

d	l	m	m	j	v	s
				1	2	3
4	5	6	7	8	9	10
11	12	13	14	15	16	17
18	19	20	21	22	23	24
25	26	27	28	29	30	

octubre

d	l	m	m	j	v	s
						1
2	3	4	5	6	7	8
9	10	11	12	13	14	15
16	17	18	19	20	21	22
23	24	25	26	27	28	29
30	31					

noviembre

d	l	m	m	j	v	s
		1	2	3	4	5
6	7	8	9	10	11	12
13	14	15	16	17	18	19
20	21	22	23	24	25	26
27	28	29	30			

diciembre

d	l	m	m	j	v	s
				1	2	3
4	5	6	7	8	9	10
11	12	13	14	15	16	17
18	19	20	21	22	23	24
25	26	27	28	29	30	31

Online Content at **connectED.mcgraw-hill.com**

Yy

yesterday

yesterday   today

			April			
Sunday	Monday	Tuesday	Wednesday	Thursday	Friday	Saturday
		1	2	3	4	5
6	7	8	9	10	11	12
13	14	15	16	17	18	19
20	21	22	23	24	25	26
27	28	29	30			

ayer

ayer   hoy

			abril			
domingo	lunes	martes	miércoles	jueves	viernes	sábado
		1	2	3	4	5
6	7	8	9	10	11	12
13	14	15	16	17	18	19
20	21	22	23	24	25	26
27	28	29	30			

Zz

zero

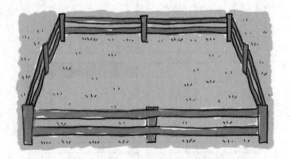

cero

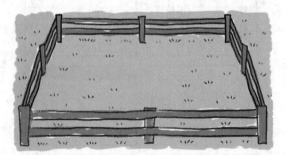

Work Mat 7: Sorting Mat/T-Chart

Work Mat 8: Two-Part Mat

Work Mat 8: Two-Part Mat